PAST & PRESENT

ARNHEM
MARKET GARDEN 1944

Simon Forty and Tom Timmermans

Casemate
PHILADELPHIA & OXFORD

Published in the United States of America and Great Britain in 2017
by CASEMATE PUBLISHERS
1950 Lawrence Road, Havertown, PA 19083
and 10 Hythe Bridge Street, Oxford, OX1 2EW

ISBN-13: 978-1-61200-540-9

Produced by Greene Media Ltd.

Cataloging-in-publication data is available from the Library of Congress
and the British Library.

10 9 8 7 6 5 4 3 2 1

Printed and bound in China
For a complete list of Casemate titles please contact:
CASEMATE PUBLISHERS (US)
Telephone (610) 853-9131, Fax (610) 853-9146
E-mail: casemate@casematepublishers.com

CASEMATE PUBLISHERS (UK)
Telephone (01865) 241249, Fax (01865) 794449
E-mail: casemate-uk@casematepublishers.co.uk

Acknowledgments

Most of the contemporary photos are from BattlefieldHistorian.com, to
whom grateful thanks, NARA College Park, MD, and the George Forty
Library. The modern photos are by Tom Timmermans; other credits are
noted on the photographs. If anyone is missing or incorrectly credited,
apologies: please notify the authors through the publishers. Thanks to Linda
Dawes and Dr Edward Bujak of Harlaxton Manor for their help.

I'd like to thank in particular Tom Timmermans and Battledetective.com
for the past and present photos. Other thanks are due to Leo Marriott (aerial
photos), Barry van Veen of STIWOT (Traces of War is undoubtedly the best
way of planning an intinerary around the key sites), Mark Franklin (maps),
Ian Hughes (design), Stephen Smith (editorial work), Richard Wood and the
military cyclists (particularly Peter Anderson) for photos and enthusiasm.

In particular, any author on this subject owes a huge debt to After the Battle
and Karel Margry's brilliant pair of books, *Operation Market Garden Then and
Now*, and to the equally brilliant Pegasus Archive, whose section on Arnhem
contains much primary source material (unit war diaries) and a comprehensive
coverage of the battle. Martin Middlebrook's *Arnhem 1944* and Robert
Kershaw's *It Never Snows in September* provided much of the information used
to create the maps.

Previous page:
The Soldier with
the Flower Girl
Memorial was
unveiled outside the
Hartenstein Museum
in 2011.

Right:
Reenactors parked
outside the St.
Elisabeth Hospital.

Contents

Introduction

Above:
The Airborne badge depicts Bellerophon riding Pegasus. It was designed by Edward Seago, from a suggestion by the famous novelist Daphne du Maurier, the wife of Maj-Gen "Boy" Browning, commander of 1st Airborne. It was worn by all British Airborne troops.

Below:
American Lt-Gen Lewis Hyde Brereton (1890–1967) commanded the newly constituted First Allied Airborne Army with the British Lt-Gen Frederick Browning (1896–1965) his deputy.

We've all seen the movie, and as is so typical of any Hollywood blockbuster, it has defined the battle for many people. Relying heavily on a few vocal sources, it had to find a villain and he was provided by someone who could not defend himself: "Boy" Browning was dead. The willful supression of the intelligence of two "crack" SS-Panzer Divisions; the wrong crystals taken for the radios; the tardiness of the British armor at Nijmegen—in the end, the power of mass-market cinema has helped obscure the reasons for the failure of Operation Market Garden, a daring attempt to use the First Allied Airborne Army to end the war by Christmas.

In fact, as Martin Middlebrook so succinctly outlines in the final section of his book on the Airborne battle, the real reasons behind the failure of the plan have been identified, with—on the Allied side—the air plan being pivotal. The inflexibility of Maj-Gen. Paul Williams, commander of IX Troop Carrier Command, meant that the British troops were dropped over three days, miles from the bridge, without the sort of coup de main operation that had proved so successful on the Orne on D-Day. The German assessment of the battle suggested the Allies' chief mistake was this protracted, three-day landing. Added to this, no ground-support missions could be flown while the air drops were taking place, and that included Flak suppression.

It seems incredible but the man in charge of close air support over the Arnhem area—the commander of 2TAF, AM Sir Arthur Coningham—was not invited to attend any of planning meetings until the day before the operation, when the weather was too bad for him to attend. This meant that the first ground-support mission flown directly in aid of the Paras in the front line was not till September 24.

The ground-support missions weren't helped by the fact that the US Air Support Signals Team, the 306th Fighter Control Squadron which accompanied the Paras in four Wacos from Manston, had been given the wrong crystals and were unable to make contact with their air assets. Apart from this, the problems with radio communication, so often alluded to (not least by Maj-Gen. Roy Urquhart), were not down to poor or inadequate equipment but basic procedural errors.

Dropped so far from the bridge over a three-day period, the division needed to protect the LZs and DZs against enemy interference, thus reducing significantly the number of troops heading for the bridge and speed with which they traveled. This, in turn, meant that the immediate German response to the airborne operation—ad hoc groupings of whatever troops were to hand, not the reaction of heavily armed "crack" SS troops, which were anyway at only 20–30 percent of their established strengths—was remarkably effective, in a way that would have almost certainly not been the case in the face of larger numbers of troops. Without the blocking action of KG Krafft, the Airborne Recce Sqn jeeps may well have reached the bridge; without the actions of KG Spindler on September 17–18, Frost's 2nd Para Battalion would have received vital reinforcements and ammunition.

Robert Kershaw's brace of Arnhem books supply a lucid account of the German side of the operation and highlight this speedy response to the arrival of 1st Airborne. As he says in his assessment, "too often Allied historians have tended to blame mistakes rather than effective countermeasures." The speed of the German reaction was crucial to their success. On September 7 six British battalions landed near Arnhem.

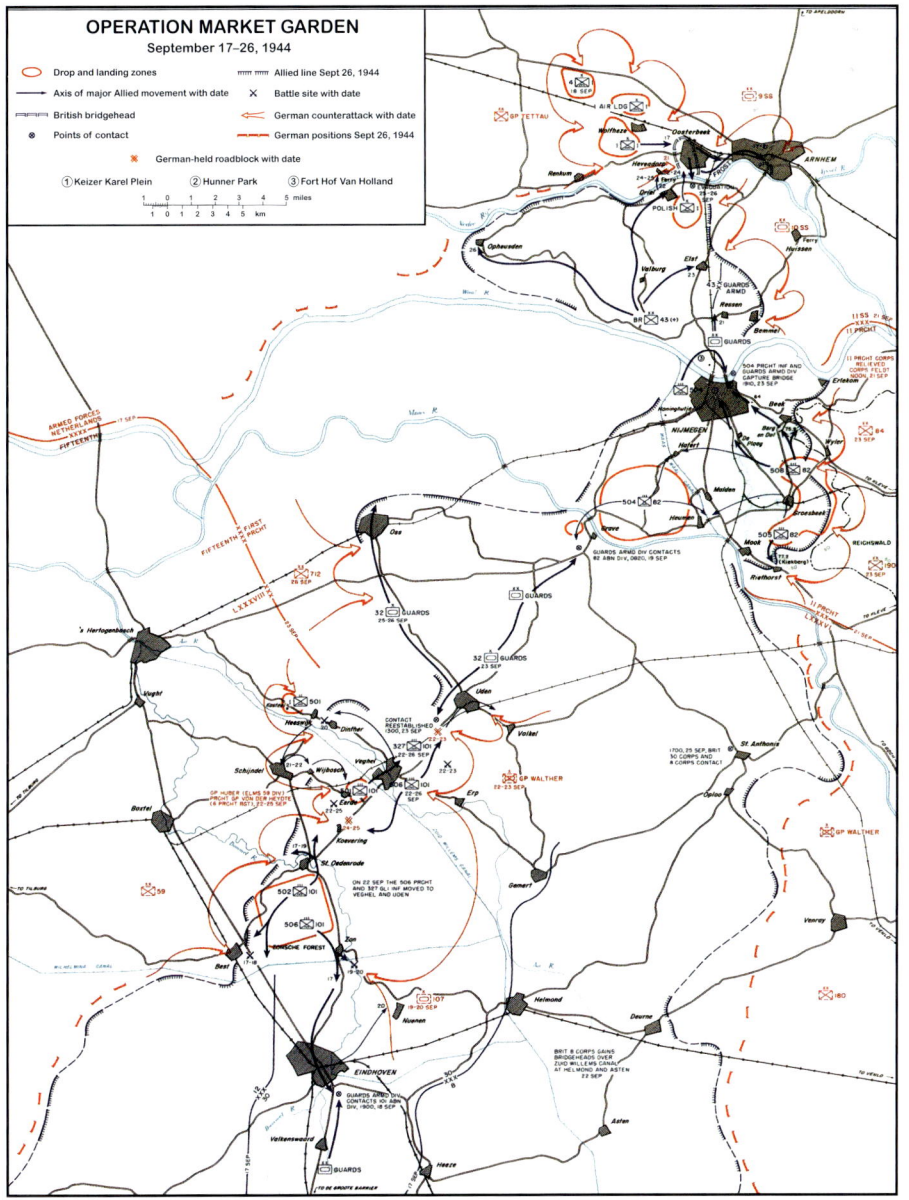

OPERATION MARKET GARDEN
September 17–26, 1944

⬭ Drop and landing zones   ▥▥ Allied line Sept 26, 1944

→ Axis of major Allied movement with date   ✕ Battle site with date

⬭▭ British bridgehead   ⇐ German counterattack with date

⊛ Points of contact   ▬▬▬ German positions Sept 26, 1944

     ✕ German-held roadblock with date

① Keizer Karel Plein   ② Hunner Park   ③ Fort Hof Van Holland

1 0 1 2 3 4 5 miles
1 0 1 2 3 4 5 km

By midnight—less than twelve hours after the first Para had dropped—there were between ten and eleven German battalions opposing them, their troop movements having taken place without harassment by Allied close air support. By September 20 the Germans had a three-to-one advantage. They had, as Kershaw points out, won the reinforcement battle.

Part of the reason for this was that the Germans were able to divert reinforcements heading towards the US First Army sector—for example the Tiger IIs of 2./sPzAbt 506 destined for Aachen. Montgomery's plan had called for a thrust along the Aachen–Cologne corridor at the same time as Market Garden. That thrust didn't reach the anticipated levels.

The end result was the annihilation of British 1st Airborne Division, whose staggering bravery resonates to this day, as does that of the crews of the aircraft, the air dispatchers, the civilians of Arnhem, and the German defenders.

Right:

There were many heroes during the fighting and five VCs were awarded. Lance Sergeant "Jack" Baskeyfield received one of them.

A member of the South Staffordshire Regiment, he was in charge of a six-pounder AT gun at Oosterbeek. On September 20, his gun knocked out two tanks and at least one SP gun. Wounded in the leg, he manned his gun alone as further attacks were launched and driven off. When his gun was knocked out, he crawled under fire to another gun which he again manned single-handed, engaging and halting an SP gun, before a supporting enemy tank knocked out his gun and killed him. This statue of Jack Baskeyfield was unveiled on November 17, 1996, in Stoke-on-Trent. *Steven Whyte/WikiCommons (CC BY-SA 3.0)*

Right:

A wall relief at the Berenkuil (Bearpit) roundabout just north of the Arnhem road bridge.

Below:

Coup de main: a bridge taken and held—Pegasus Bridge at Benouville.

ORDER OF BATTLE OF THE 1st BRITISH AIRBORNE CORPS

Commander: Lt-Gen. Frederick Browning

1st AIRBORNE DIVISION

(CO: Maj-Gen. Roy Urquhart)
  Div HQ and Defence Pl

1st Parachute Bde

Brigade HQ and Defence Pl
  CO: Brig. Gerald Lathbury
1st Para Bn (Lt-Col. David Dobie)
  HQ/Support, R, S, and T Coys
2nd Para Bn (Lt-Col. John Frost)
  HQ, Support, A, B, and C Coys
3rd Para Bn (Lt-Col. John Fitch)
  HQ/Support, A, B, and C Coys
1st Airlanding AT Bty, RA (Maj. Bill
  Arnold) A, B, C, D, P, and Z
  Troops
1st Para Sqn, RE (Maj. Douglas
  Murray) A, B, and C Troops
16 Para Fd Amb, RAMC (Lt-Col. E.
  Townsend)

4th Parachute Bde

Brigade HQ and Defence Pl
  CO: Brig. John Hackett
10th Para Bn (Lt-Col. Ken Smyth)
  HQ, Support, A, B, and D Coys
11th Para Bn (Lt-Col. George Lea)
  HQ, Support, A, B, and C Coys
156th Para Bn (Lt-Col. Sir Richard
  Des Voeux) HQ, Support, A, B,
  and C Coys
2nd (Oban) Airlanding AT Bty, RA (Maj.
  A. Haynes) E, H, G, H, and X
  Troops
4th Para Sqn, RE (Maj. Aeneas
  Perkins)
  1, 2, and 3 Troops
133 Para Fd Amb, RAMC (Lt-Col. W.
  Alford)

1st Airlanding Bde

Brigade HQ and Defence Pl
  CO: Brig. Philip Hicks
1st Bn The Border Regt (Lt-Col.
  Tommy Haddon) HQ, Support, A,
  B, C, and D Coys
2nd Bn The South Staffordshire Regt
  (Lt-Col. Derek McCardie) HQ,
  Support, A, B, C, and D Coys

*7th (Galloway) Bn The King's Own
  Scottish Borderers* (Lt-Col. Robert
  Payton-Reid) HQ, Support, A, B,
  C, and D Coys
133 Airlanding Fd Amb, RAMC (Lt-Col.
  Arthur Marrable)

Div Units

1st Airlanding Lt Regt, RA (Lt-Col.
  "Sheriff" Thompson) 1st, 2nd, and
  3rd Airlanding Lt Btys (each with
  3 Troops)
*1 Forward (Airborne) Observation Unit,
  RA* (Maj. Denys Wight-Boycott)
1st AB Div Signals (Lt-Col. Tom
  Stephenson) No.1 Coy (Div HQ
  Signals of 2 Sects); No.2 Coy
  (Bde and Arty Signals of 5 Sects
  each with unit HQs); 1st Polish
  Ind Parachute Bde Group; 1st AB
  Corps Signals (at Nijmegen)
1st AB Recce Sqn (Maj. Freddie
  Gough) HQ, Support, A, C, and D
  Troops
21st Ind Para Coy (Maj. "Boy"
  Wilson) 1, 2, and 3 Platoons
9th (AB) Fd Coy, RE (Maj. John
  Winchester) 1, 2, and 3 Platoons
261 (AB) Field Park Coy, RE (Lt. W.
  Skinner
250 (AB) Lt Composite Coy, RASC—
  Para Platoons and Jeep Sects
  allocated to 1st Para, 4th Para,
  1st Airlanding Bdes
*1st (AB) Div Ordnance Field Park,
  RAOC* (Capt. Bill Chidgey)
1st (AB) Div Workshops, REME
1st (AB) Div Provost Coy, CMP (Capt.
  Bill Gray) 4 Sects
*89th (Para) Field Security Section,
  Intelligence Corps* (Capt. J. Killick)

The Glider Pilot Regt

CO: Col. George Chatterton
No.1 Wing (Lt-Col. Iain Murray) A, B,
  D, and G Sqns
No.2 Wing (Lt-Col. John Place) C, E,
  and F Sqns

Other Troops

Dutch Liaison Mission: No.2 (Dutch)
  Troop of No.10 (Inter-Allied)
  Commando
*6080 and 6341 Light Warning Units,
  RAF*
US Air Support Signals Team: 306th
  Fighter Control Sqn

1st (POLISH) INDEPENDENT PARA BRIGADE GROUP

(CO: Maj-Gen. Stanislaw
  Sosabowski)
1st Battalion (Maj. M. Tonn)
2nd Battalion (Maj. W. Ploszewski)
3rd Battalion (Capt. W. Sobocinski)
AT Battery (Capt. J. Wardzala)
Engineer Company (Capt. P.
  Budziszewski)
Signals Company (Capt. J. Burzawa)
Medical Company (Lt. J. Mozdzierz)
Transport and Supply Company
  (Capt. A. Siudzinski)
Light Artillery Battery (Maj. J. Bielecki)

Right:
The Polish eagle
emblem of the
1st Ind Para
Bde Group.

Battle Casualties

Considering the intensity of the fighting and its duration, it's unsurprising that 1st Airborne's casualty figures make sorry reading: over a thousand killed in action or dead of wounds. By the far the largest casualty figure is that of the captured or missing—many of them wounded—most of whom entered captivity when the Oosterbeek Perimeter was overrun. 1st Airborne in effect ceased to exist as a unit and 4th Para Bde was disbanded soon after. The division went on to supervise the surrender of German forces in Norway (Operation Doomsday) before it, too, was disbanded in 1945.

	KIA or died of wounds	Captured or missing	Evacuated	Total
1st Airborne	1,174	5,903	1,892	8,969
Glider Pilot Regiment	219	511	532	1,262
Polish Brigade	92	111	1,486	1,689
Total	1,485	6,525	3,910	11,920

Opposite:
Willemsplein square sports a different viaduct but the scene is appreciably the same today as it was on September 19 when Sturmgeschütz-Brigade 280 arrived in the city. Having spent some time refitting in Denmark, they were diverted to Arnhem while on their way to Aachen. The ten StuGs—seven StuG IIIs and three StuG42s— were commanded by Major Kurt Kühme and turned the battle in the Germans' favor, initially three each being allocated to KGs Harder, Moeller, and von Allworden and used in the direct fire support role.

This page:
Today, Arnhem's railroad station seems to have gotten bigger! Lion route—that followed by Lt-Col John Frost and his men— passed close to the river at this point.

9

Opposite, Center left:

The new Battle of Arnhem Information Center near the John Frost Bridge reopened on March 30, 2017.

Opposite, Center right:

The oldest of the memorials, created in 1945, is at Heelsum. It's been moved but the 6pdr is still an imposing presence.

Opposite, Below:

Jacobus Groenwoud was part of Jedburgh Team "Claude" and was killed while leaving Frost's perimeter trying to reach British lines. The Jacobus Groenewoud Plantsoen remembers the only Dutch officer to die during the fighting. The Canadian 25-pounder wasn't used in the battle.

Map and key below:

Today there are many museums, memorials, and reminders of the battle to visit—particularly in the Arnhem/Oosterbeek area. This map identifies the key locations.

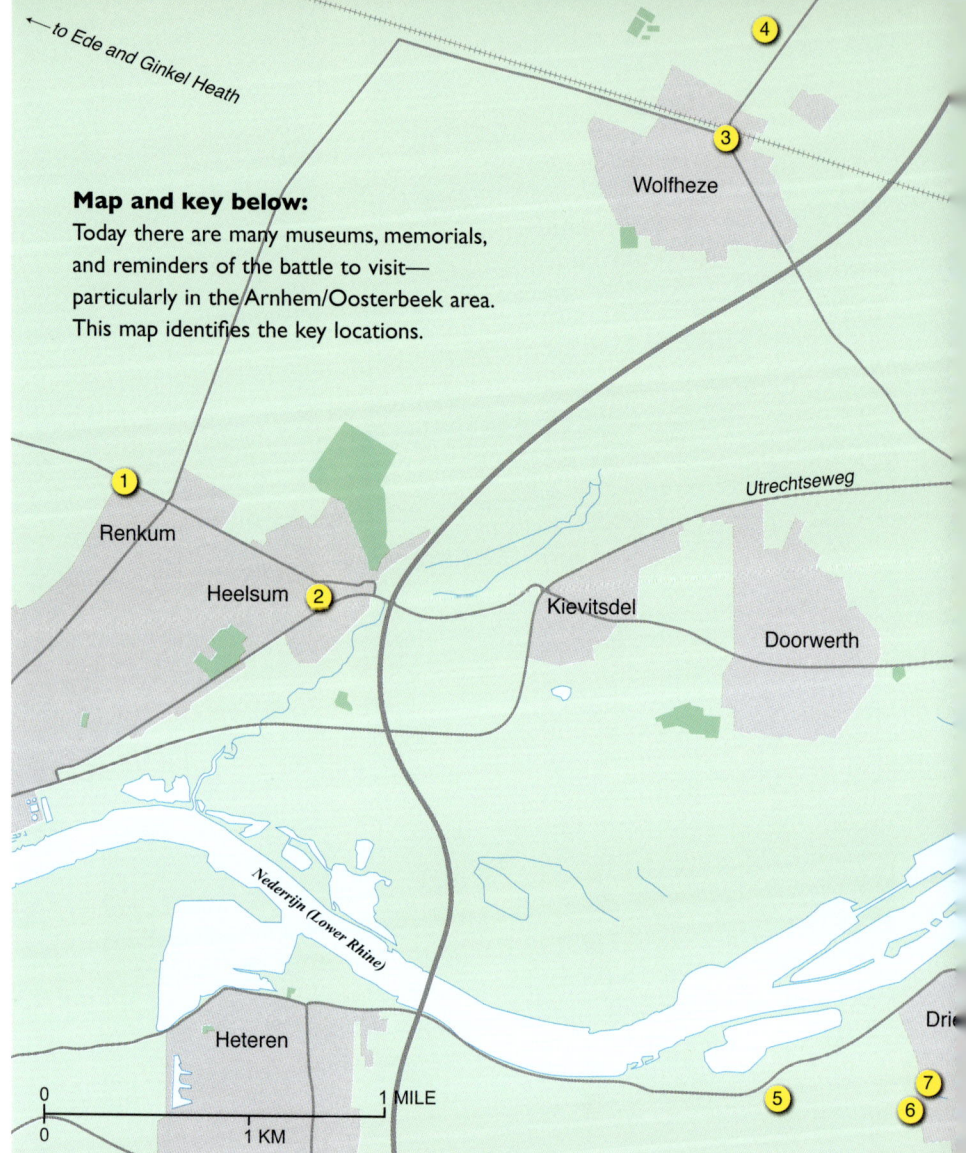

1 Eethuys Airborne restaurant (Telefoonweg, Renkum)
2 Airborne Memorial (Bennekomseweg, Heelsum)
3 Airborne Memorial (Wolfhezerweg, Wolfheze)
4 Glider Memorial (Wolfhezerweg, Wolfheze
5 Hampshire Regt Memorial (Drielse Rijndijk, Driel)
6 Sosabowski Memorial (Polenplein, Driel)
7 Plaque 5th Bn DCLI (Kerkstraat, Driel)
8 Battle Marker 4* (Valkenburglaan, Oosterbeek)
9 Dorset Terrace, Westerbouwing Restaurant (Westerbouwing, Oosterbeek)
10 Airborne Memorial (Utrechtseweg, Oosterbeek)
11 Hartenstein Airborne Museum and memorials (Utrechtseweg, Oosterbeek)
12 1st AB Recce Sqn Memorial* (Oranjeweg, Oosterbeek)
13 21st Ind Para Coy Memorial (Utrechtseweg/Stationsweg, Oosterbeek)
14 Battle Marker 1* (Hotel Tafelberg, Pietersbergseweg, Oosterbeek)
15 Battle Marker 5* Lonsdale Church (Benedendorpsweg, Oosterbeek)
16 Memorial Operation Berlin (Polderweg, Oosterbeek)
17 Memorial RE and RCanE (Drielse Rijndijk)
18 Battle Marker 3* (Dreijenseweg, Oosterbeek)
19 Commonwealth War Cemetery (Van Limburg Stirumweg, Oosterbeek)
20 Air Despatchers Memorial (Van Limburg Stirumweg, Oosterbeek)
21 Battle Marker 2* (Utrechtseweg, Arnhem)
22 September 1944 Memorial (Nassaustraat, Arnhem)
23 Battle Marker 6* (St. Elisabeth's Hospital, Utrechtseweg, Arnhem)
24 Airborne House (Utrechtseweg, Arnhem)
25 Battle Marker 9* (Utrechtseweg, Arnhem)
26 Battle marker 7* (Boterdijk, Arnhem)
27 Plaque HQ John Frost (Prinsenhof, Arnhem)
28 St. Eusebius' Church (Kerkplein, Arnhem)
29 Airborne at the Bridge (Rijnkade, Arnhem)
30 Airborne Monument (Berenkuil, Arnhem)
31 Battle Marker 8* (Eusebiusbuitensingel, Arnhem)
32 John Frost Monument (Nijmeegseweg, Arnhem) and Bridge

* See pp12–13

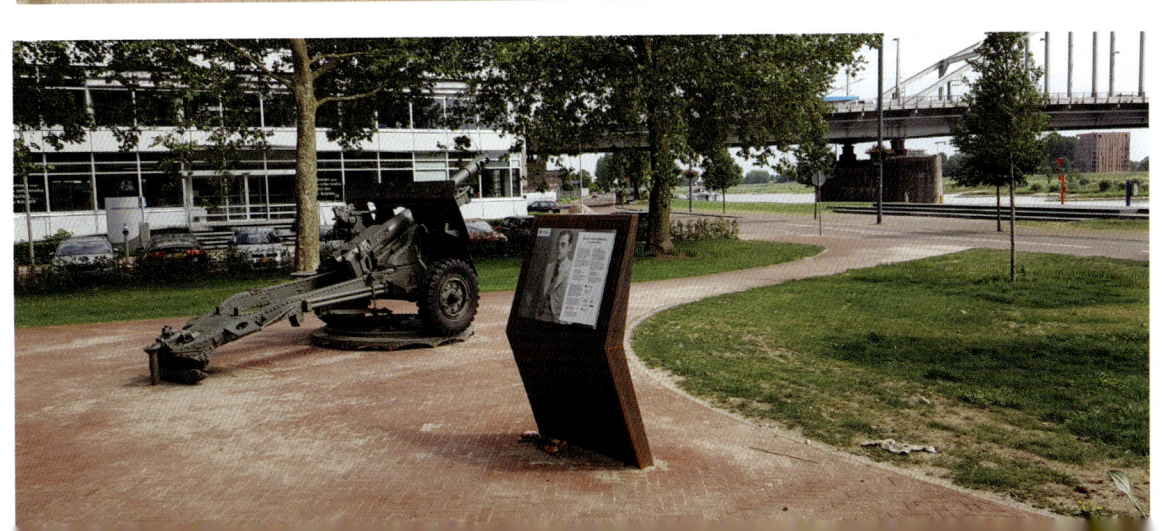

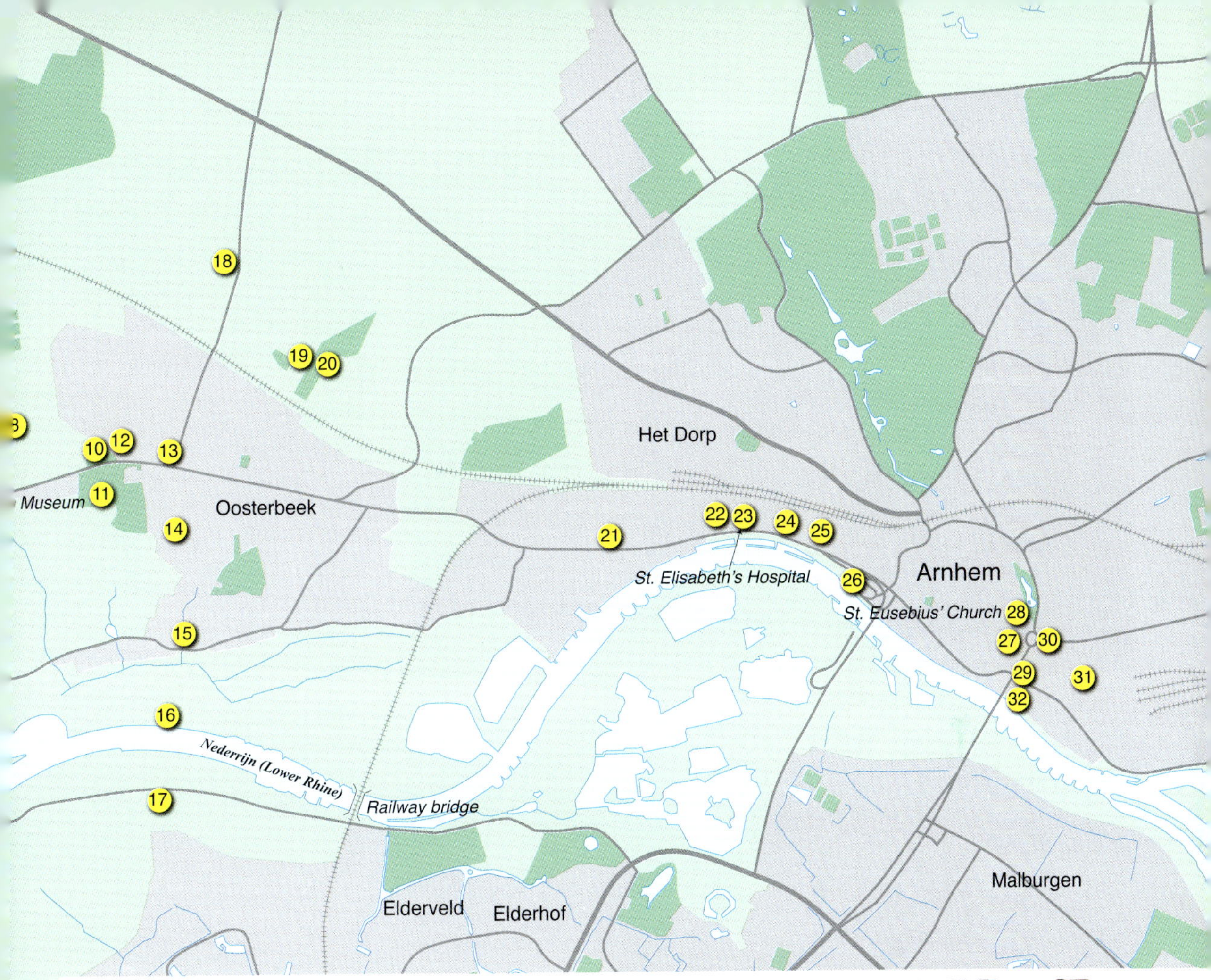
18
19
20
8
10
12
13
Museum
11
Oosterbeek
14
15
16
Nederrijn (Lower Rhine)
17
Railway bridge
Het Dorp
21
22
23
24
25
St. Elisabeth's Hospital
26
Arnhem
St. Eusebius' Church
28
27
30
29
31
32
Elderveld
Elderhof
Malburgen

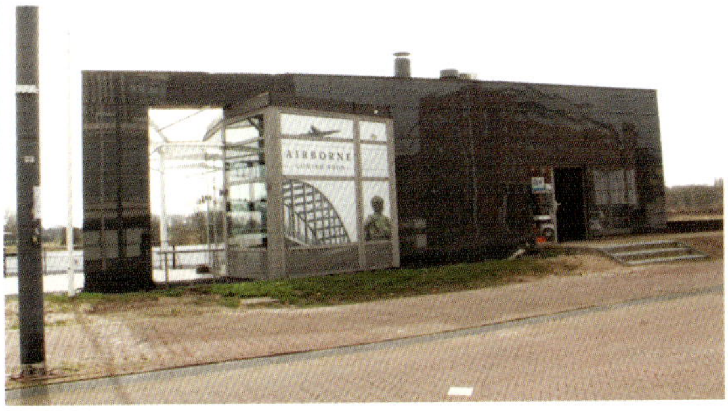
AIRBORNE

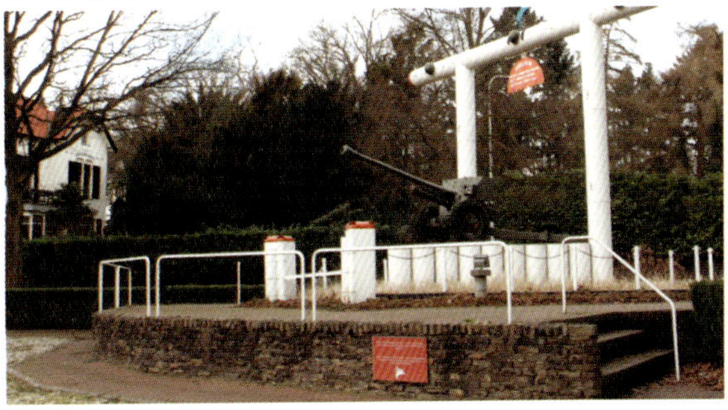

The Markers

There are a number of Airborne Commemorative Markers around the Arnhem area identifying key points in the battle. There were originally eight, but two other similar markers are included.

1 The Tafelberg was used as a medical dressing station under Maj. Guy Rigby Jones, one of two surgeons with 181 Field Ambulance Medical Team, between September 19 and 24.

2 Sited in the Old Pastor's Garden, Utrechtseweg, this marker remembers 10 (Sussex) Para Bn who fought "to virtual extinction."

3 The Dreijenseweg marker identifies where 156 Para Bn was halted on September 19. Arriving with the second lift on the 18th it suffered heavy casualties. This was the extreme NE corner of the Oosterbeek perimeter.

4 The Hollow marker remembers the bravery of Brig. "Shan" Hackett and 4 Para Bde who held the area until, pressed by tanks and flamethrowers, they escaped to continue the fight in Oosterbeek.

5 Old Lonsdale Church—so called because it was here that Maj. Lonsdale reorganized the survivors of the fighting at St. Elisabeth Hospital.

6 St. Elisabeth Hospital is now apartments. The marker commemorates 16th Para Fd Ambulance's time there. They landed with the First Lift on

September 17, reached St. Elisabeth by 20:00 and immediately started treating casualties. From September 19 the hospital was in German hands. Most of the medics went on to serve in the Airborne Hospital set up in Apeldoorn.

7 This remembers the final attempt of 1 and 3 Para Bns to break through to the bridge on September 19 before retreating to Oosterbeek.

8 The Van Limburg Stirum School was held by 1st Para Sqn RE and C Coy, 3rd Para Bn until attacked on the morning of the 20th by a tank and an SP. One shell set the roof ablaze and wounded Maj. "Pongo" Lewis, the officer commanding. The remaining troops were forced to surrender.

9 This marker remembers the last attempt by 2nd South Staffs and 11th Para Bn to break through to the bridge on September 19. After this the survivors moved into the Oosterbeek Perimeter.

10 This marker is dedicated to the men of the 1st Airborne Recce Sqn, which lost 29 men during the operation with a further 140 becoming PoWs.

Into Battle

Below right:

The 61st Troop Carrier Group flew from Barkston— this September 17 photograph shows paras alongside the aircraft waiting to emplane.

Opposite, Above:

Church Army Mobile Canteens were a welcome source not only of tea and food but other necessities from writing materials to tooth-cleaning gear.

Opposite, Below:

The routes to the landing fields of Ginkel Heath. British 1st Airborne units were delivered by British and American troop carriers flying from Blakehill Farm, Broadwell, Down Ampney, Fairford, Harwell, Keevil, Manston, and Tarrant Rushton.

AT 09:45 ON SUNDAY, September 17, the aircraft carrying 1st Airborne Division took off from England. The first day's lift used nearly 500 transport aircraft: twelve of them carrying the Pathfinders—186 officers and men of the 21st Independent Para Coy. Their mission was to deploy the Eureka beacons, mark out the landing areas, and defend them until the first wave arrived.

Everything went according to plan. At 12:40 the Stirlings—the only RAF aircraft used to carry 1st Airborne on the first day—dropped their cargo and returned to Fairford. The Pathfinders marked out the landing grounds—LZ-S and LZ-Z, and DZ-X—and waited. The small number of Germans they came across either ran off or were taken prisoner.

The first to arrive were the gliders of 1st Airlanding Bde on LZ-S at around 13:00. Some 134 landed (152 Horsa and a Hamilcar had been planned). They were to be followed by 156 Horsas and 11 Hamilcars. In fact, at 13:20 the first of 150 gliders landed on LZ-Z including four Wacos carrying the US Air Support Signals Team and Maj-Gen. Urquhart's HQ. The 320 tug and glider combinations

had left from Tarrant Rushton, Keevil, Broadwell, Blakehill Farm, Manston, Down Ampney, Fairford, and Harwell— the airfields of 38 and 46 Groups, RAF. There had been a few failures—seventeen gliders did not reach Arnhem, including the loss of half of No.1 Platoon, 9th Field Company, RE when their Horsa broke up (see photo p.16).

It wasn't long before the Paras started dropping on DZ-X from 143 C-47 Dakotas of 314th and 61st Troop Carrier Groups that had left from US bases at Barkston Heath and Spanhoe. Slightly later than planned, at 13:50, 2,278 men dropped.

When compared to the drops on Sicily and Normandy, the delivery of the first lift on September 17 was almost miraculously good. There had been few casualties, the divisional CO and his headquarters had landed successfully, as had the 38 gliders of the Headquarters of the 1st British Airborne Corps at Nijmegen—although the need for this has been much debated since. The landings had been virtually unopposed, complete surprise had been achieved ... but the second lift was twenty hours away.

continued on page 18

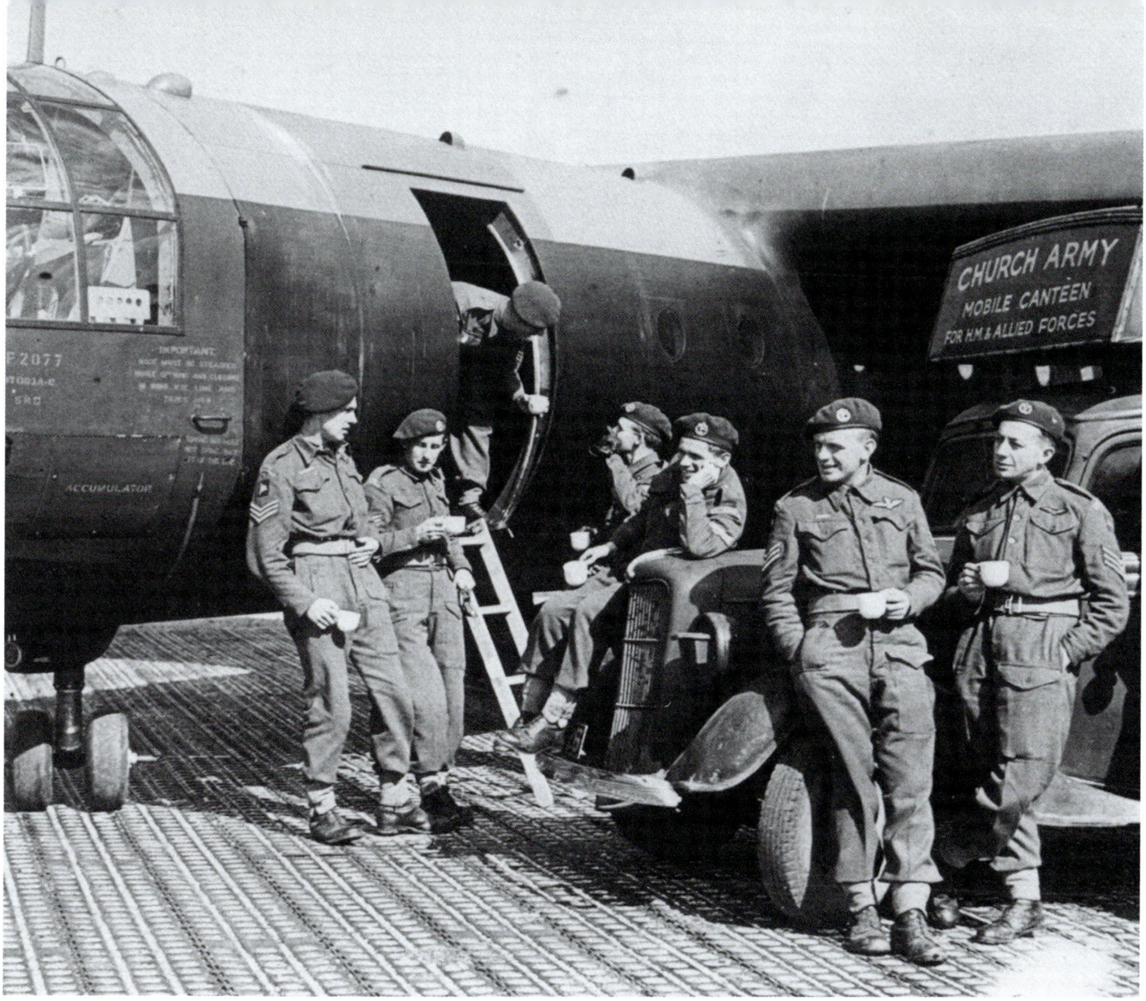

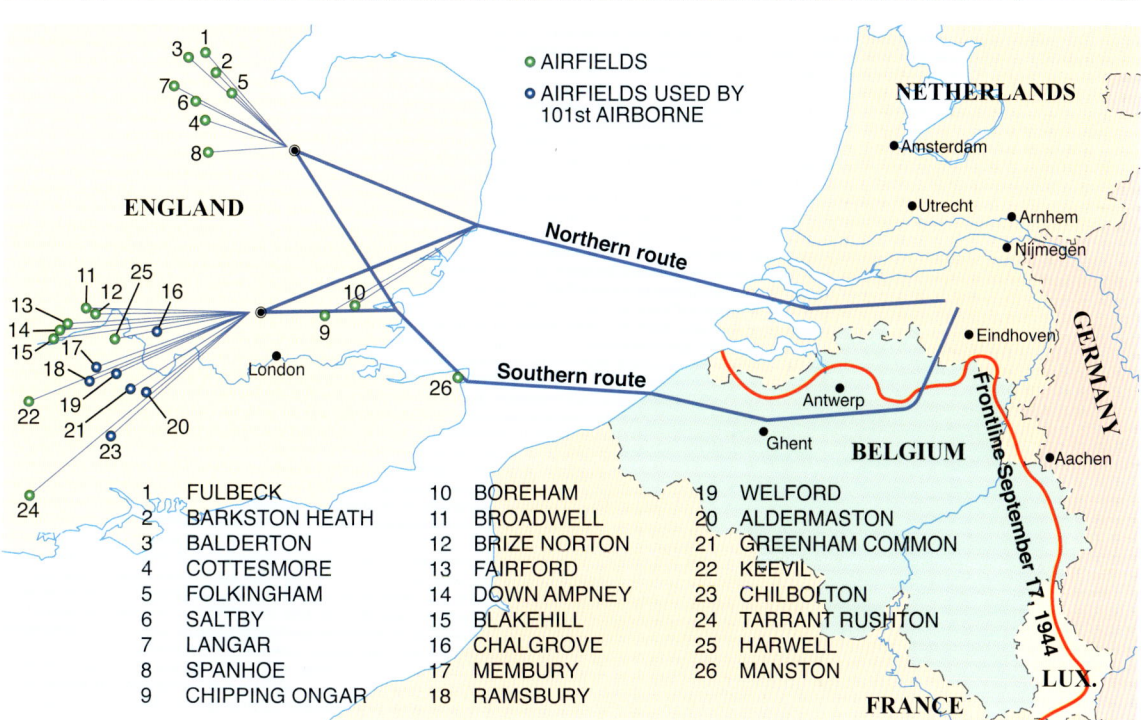

CHURCH ARMY
MOBILE CANTEEN
FOR H.M. & ALLIED FORCES

AIRFIELDS
AIRFIELDS USED BY 101st AIRBORNE
ENGLAND
NETHERLANDS
BELGIUM
GERMANY
FRANCE
LUX.
Amsterdam
Utrecht
Arnhem
Nijmegen
Eindhoven
Antwerp
Ghent
Aachen
London
Northern route
Southern route
Frontline September 17, 1944
1 FULBECK
2 BARKSTON HEATH
3 BALDERTON
4 COTTESMORE
5 FOLKINGHAM
6 SALTBY
7 LANGAR
8 SPANHOE
9 CHIPPING ONGAR
10 BOREHAM
11 BROADWELL
12 BRIZE NORTON
13 FAIRFORD
14 DOWN AMPNEY
15 BLAKEHILL
16 CHALGROVE
17 MEMBURY
18 RAMSBURY
19 WELFORD
20 ALDERMASTON
21 GREENHAM COMMON
22 KEEVIL
23 CHILBOLTON
24 TARRANT RUSHTON
25 HARWELL
26 MANSTON

1. THIS STONE IS DEDICATED TO THE MEN WHO LEFT WOODHALL SPA TO FIGHT IN THE BATTLE OF ARNHEM

TO THOSE WHO RETURNED AND TO THOSE WHO DID NOT

1st AIRLANDING BRIGADE

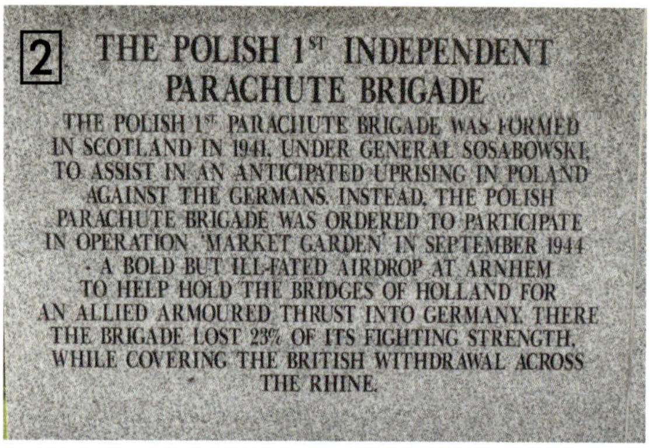

1944

2. **THE POLISH 1ST INDEPENDENT PARACHUTE BRIGADE**

THE POLISH 1ST PARACHUTE BRIGADE WAS FORMED IN SCOTLAND IN 1941, UNDER GENERAL SOSABOWSKI, TO ASSIST IN AN ANTICIPATED UPRISING IN POLAND AGAINST THE GERMANS. INSTEAD, THE POLISH PARACHUTE BRIGADE WAS ORDERED TO PARTICIPATE IN OPERATION 'MARKET GARDEN' IN SEPTEMBER 1944 - A BOLD BUT ILL-FATED AIRDROP AT ARNHEM TO HELP HOLD THE BRIDGES OF HOLLAND FOR AN ALLIED ARMOURED THRUST INTO GERMANY. THERE THE BRIGADE LOST 23% OF ITS FIGHTING STRENGTH, WHILE COVERING THE BRITISH WITHDRAWAL ACROSS THE RHINE.

3.

4.

There are a number of memorials in the UK to the airborne forces who fought at Arnhem: this is a small selection.

1 "This stone is dedicated to the men who left Woodhall Spa to fight in the battle of Arnhem. To those who returned and to those who did not. 1st Airlanding Brigade, 1944."

2 A plaque on the Polish Armed Forces Memorial at the National Memorial Arboretum in Staffordshire.

3 MLDE engineered "Pegasus" at the National Memorial Arboretum, combining bronze sculptures by Charlie Langton and Mark Jackson.

4 At Double Hills, just outside Paulton, Somerset, "This memorial is raised to the memory of the 2 glider pilots and 21 men of the 9th Airborne Field Company, Royal Engineers who were killed when their Airspeed Horsa glider No. RJ113 of D Squadron Glider Pilot Regiment crashed in this field on Sunday September 17th 1944 en-route for Arnhem Operation Market Garden. The glider took off from RAF Keevil, Wiltshire towed by a Short Stirling bomber No. LK148 of 299 Squadron. All 23 men were buried at Weston-super-Mare cemetery with full military honours." The names of the dead are inscribed below.

Right: Today part of the University of Evansville, Harlaxton Manor in Lincolnshire housed elements of 1st Airborne Division during the war including: 1st Forward (Airborne) Observation Unit, RA, who left graffiti in the refractory (**6**); 253rd Airborne Composite Coy, RASC (seen at **7** outside the walled garden with King George VI in March 1944 and today, **8**); and 1st (AB) Divisional Provost Coy, CMP, two of whose number— LCpl (later Lt) "Bill" Hinchliffe and Cpl (later Maj, MBE) Roy Tyler—created the Pegasus memorial (**5**).

Photos supplied by Linda Dawes, Harlaxton College.

The Arnhem Plan: Ultimate Defensive Positions

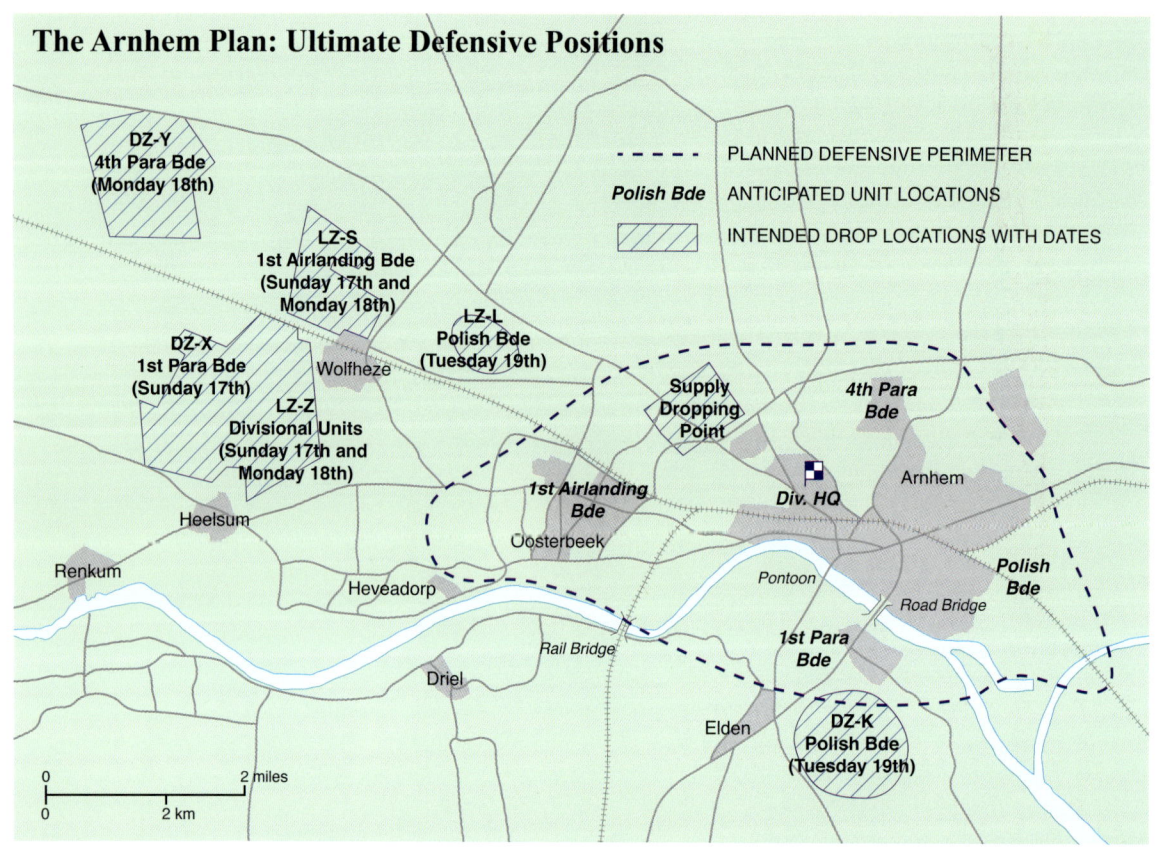

DZ-Y
4th Para Bde
(Monday 18th)

- - - - - PLANNED DEFENSIVE PERIMETER

Polish Bde ANTICIPATED UNIT LOCATIONS

INTENDED DROP LOCATIONS WITH DATES

LZ-S
1st Airlanding Bde
(Sunday 17th and
Monday 18th)

LZ-L
Polish Bde
(Tuesday 19th)

DZ-X
1st Para Bde
(Sunday 17th)

Wolfheze

LZ-Z
Divisional Units
(Sunday 17th and
Monday 18th)

Supply
Dropping
Point

*4th Para
Bde*

*1st Airlanding
Bde*

Div. HQ

Arnhem

Heelsum

Oosterbeek

Renkum

Heveadorp

Pontoon

Road Bridge

*Polish
Bde*

*1st Para
Bde*

Rail Bridge

Driel

Elden

DZ-K
Polish Bde
(Tuesday 19th)

0 ——————— 2 miles
0 ——————— 2 km

Above:

The planned dropping and landing zones. Unlike the Americans, the British landed their gliders (on LZs—landing zones) first before dropping the Paras (onto the DZs). The plan was that once landed, the troops would advance into Arnhem, establish a perimeter, and await XXX Corps' arrival. Note the Polish Bde's intended dropzone and perimeter location.

The second lift started four hours late—around 11:20—because of foggy conditions in England, although this fact couldn't be made clear to Urquhart in Arnhem because communications hadn't been established.

The second lift was not as straightforward as the first. The Germans knew that it was coming, and the KOSB who were defending the landing areas lost at least fifteen men killed in the fighting. A number of aircraft were shot down en route and their gliders didn't reach Arnhem. Just over 1,900 men of 4th Para Bde dropped on DZ-Y from 15:09 and around 30 were killed in the drop and subsequent fighting. The 126 delivery aircraft from 314th and 315th Troop Carrier Groups were flown with great bravery by their American crews and none were lost on the return journey to Saltby and Spanhoe.

Soon after, 273 gliders arrived—69 on LZ-S, the rest on LZ-X—and the remainder of the Airlanding Bde infantry arrived safely as did the main Div HQ units, ten Polish AT guns, and the rest of the divisional troops.

Less good was the resupply drop aimed for LZ-L by 31 RAF Stirlings (two others were lost bringing the number of aircraft losses to ten): less than 12 tons was retrieved.

The third lift was supposed to have taken place on Tuesday, September 19. Postponed because of bad weather in England, the Polish parachute drop did not take place until Thursday, September 21. The lack of the Polish troops would certainly play its part in the battle. However, the Polish glider drop went ahead on the 19th as planned, thirty-five Horsas taking off from Tarrant Rushton and Keevil along with eight British gliders that had had to abort on the first two lifts. Only thirty of these gliders landed.

Right:

3rd Battalion dropping. Billeted in Spalding, they emplaned at Saltby and dropped around 14:00. They would lose their CO, Lt-Col. John Fitch. Only 27 men made it back over the Rhine.

Below:

The disposition of German forces around Arnhem on September 17. Much has been made of the proximity of II. SS-Panzerkorps' two SS-Panzer Divisions (9th Hohenstaufen and 10th Frundsberg), and their part in the battle. It's worth emphasizing that the initial defense was made up of more ad-hoc units pulled together from whatever was available.

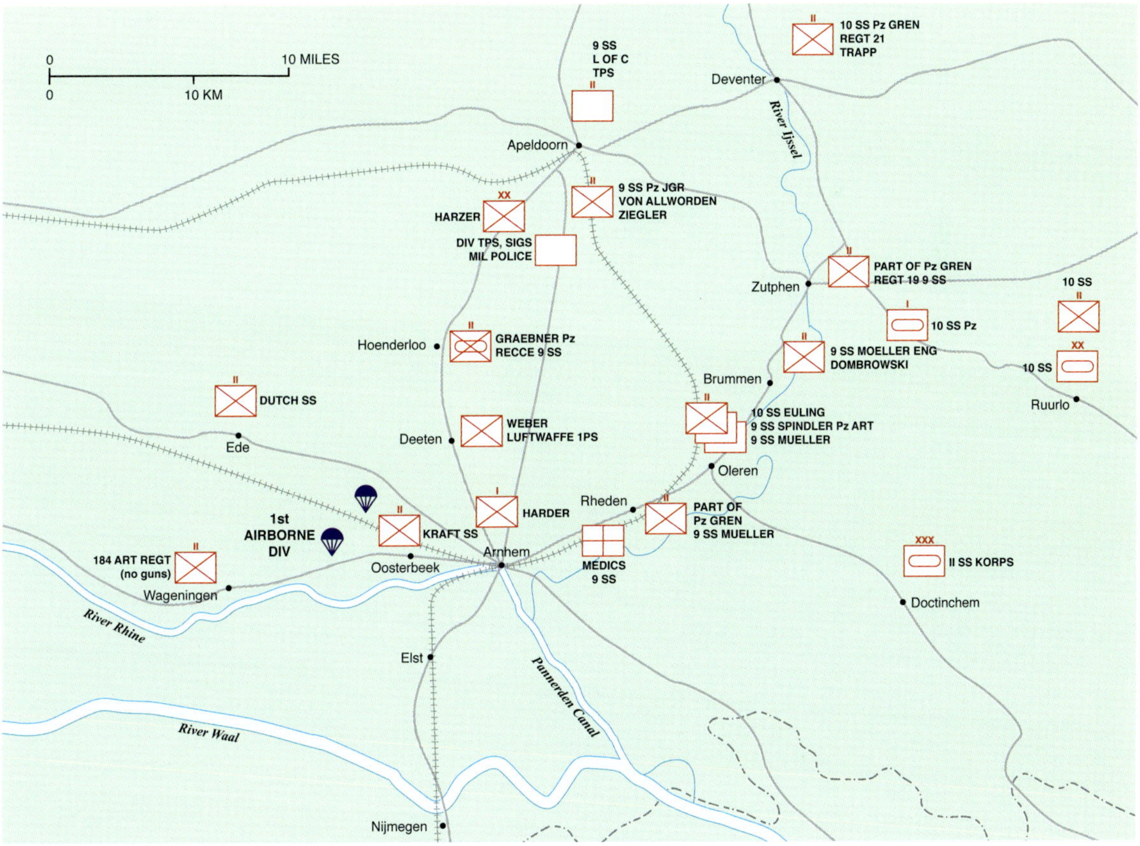

Left:
The drop on DZ–X was well-nigh perfect ... pity it was so far from the bridge.

Below:
The wide open spaces to the east of Arnhem remain largely unspoilt although crossed by the A12 motorway (**A**) from Den Haag to the German border built postwar. The landing sites are identified: DZ–X (**B**); LZ–Z (**C** and **D**); DZ–Y at Ginkel Heath (**E**); LZ–S (**F** and **G**). Wolfheze is at **H**.

Right:
The drop on LZ–Z, Renkum Heath, by 1st Bn, on September 17. Their mission was to capture the high ground north of Arnhem but, as elsewhere, the German defense proved too strong and they ended in the Oosterbeek Perimeter. Only around 100 officers and men of the nearly 550 dropped withdrew across the Rhine.

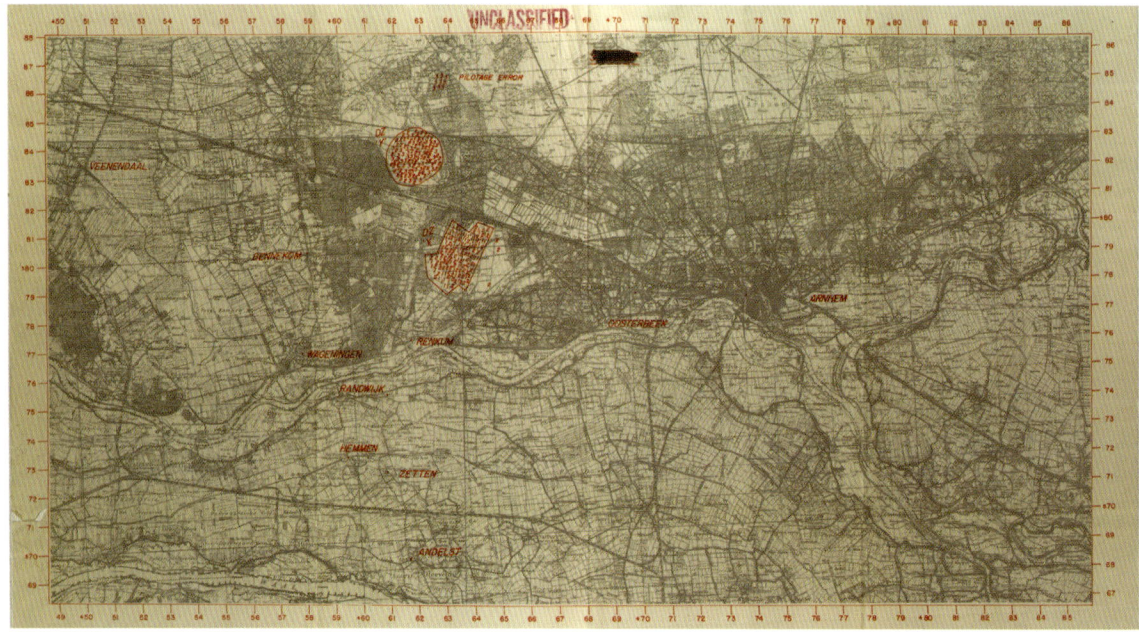

Above:

3rd Bn dropped at 14:00—accurately. Its War Diary identifies its first action coming when B Coy, the advance guard, ran into infantry and two armored cars at 17:00. Without PIATs and with their 6pdr KO'd, the action held them up for an hour.

Below:

The paradrop at Arnhem was immaculate. This graphic shows the D and D+1 drops by 52nd Wing. Trouble was that they were spread over two days, reducing the impact of surprise and allowing the German defenders to scrape together a defense.

Above:

Monument in the shape of a landing Horsa Glider opposite LZ–S, where the 1st Airlanding Bde arrived at 13:00. All but 19 of their gliders arrived, but one of those missing was that of the unlucky CO of 1st Borders, Lt-Col Tommy Haddon. The towrope on his glider broke and he relanded at Broadwell. He set off on the second lift only for his glider to be damaged by flak and land behind enemy lines. He and his men evaded capture and joined the Dorset Regt advancing toward Arnhem. When he finally reached Driel, he took part in an abortive attempt by the Dorsets to reinforce the Oosterbeek Perimeter. He and many others were captured and Haddon ended up in Oflag XIIB.

Right:

This stone remembers the landings on September 17 and 18 at Ginkel Heath, Ede, DZ–Y, where 4th Para Bde and other supporting arms landed on D+1 (September 18) the area having been secured by 7th

KOSB who landed along with the rest of the 1st Airlanding Bde on LZ–S. Ginkel Heath was the most distant of the dropzones, some 10km from the bridge.

Right:

This memorial at Ginkel Heath was erected in 1960. It bears a quote from Isiah— "They shall mount up with wings as eagles"—and the crests of the 7th KOSB, the Airborne badge, and a Pegasus.

The German Response

Below:

One of the reasons always cited for the failure of the operation is the strength of the German forces on the ground—particularly the two "crack" SS divisions of II SS-Pz Corps. Robert Kershaw in *It Never Snows in September* assesses these as 9th SS Hohenstaufen c. 2,500 men thinning out as they returned to Siegen for replenishment, and 10th SS Frundsberg with less than 3,000 men. However, they had been trained for an anti-airborne landing role, emphasizing quick reactions and junior officer and NCO decision-making and they were well placed around Arnhem to react quickly.

Opposite, Above left:

Obersturmbannführer Walter Harzer, CO of 9th SS-Pz Div Hohenstaufen, was awarded the Knight's Cross for his handling of the division at Arnhem.

THE ARRIVAL OF THE PARAS took the Germans by surprise. In the first few hours after the initial landings there is no doubt that a larger force would have been able to fight its way into Arnhem. As it was, half of the troops who landed on D-Day had to protect the landing areas for the second wave, themselves delayed by bad weather until the afternoon of D+1. By that time, the battle was strategically if not lost then severely compromised.

A key to the initial German defense was KG Krafft with just over 400 men, many of them young recruits. But the hard core of officers and NCOs were veterans and by taking up a blocking position on the railroad and Utrechtseweg, KG Krafft was able to intercept the late-starting Reconnaissance Squadron's jeeps as they sped into Arnhem to take the bridge. The Recce Sqn was stopped dead.

Elsewhere, the Germans had a stroke of luck. A glider had been shot down near Vught. In it were found a set of orders for Operation Market which were quickly taken to Genral Kurt Student.

The commander of II. SS-Pz-Korps, Gen. Bittrich, heard about the landings very quickly—at around 13:00 on D-Day—and sprang to action ordering 9th SS-Pz and 10th SS-Pz divisions to move. The former to recce Arnhem and Nijmegen before attacking the Paras in Arnhem; the latter to head for Nijmegen. The 9th SS-Pz Div alert forces were placed into KG Spindler, the CO of Hohenstaufen's Artillerie-regiment 9. He advanced into Arnhem on the evening of September 17 and set up a blocking line into which KG Krafft was subsumed.

Generalfeldmarschall Model reached Bittrich's HQ by 15:00, confirming the divisional orders, arranging the disposition of troops to counter XXX Corps and the US paratroopers, giving Student Fifteenth Army's 59th Inf Div and 107. Pz Bde which had been heading toward Aachen, and also rerouting 280th StuG-Bde, also bound for Aachen, to Arnhem. In the west, Gen. Christiansen, Armed Forces Commander, Netherlands, provided KG von Tettau.

Fast as the Germans' reactions were, they had not been fast enough to stop Frost's 2nd Para Bn reaching Arnhem Bridge. They were attacked by KGs Brinkmann and Knaust, the latter including eight PzKpfw IIIs and IVs. Savage street-fighting erupted on the morning of September 18, but 2 Para was able to beat off the attacks. It also annihilated the next attack, from the south by Graebner's Aufklärungsabteilung 9.

Left:

Obersturmbannführer Heinz Harmel (1906–2000) commanded the Frundsberg Division from April 1944. Kershaw dubs him a "wily old veteran" and he had—through a variety of means—ensured that his men and equipment had been organized carefully.

Below:

The disposition of German forces around Arnhem is shown on p. 19. This map shows the reaction of those forces to the British landings.

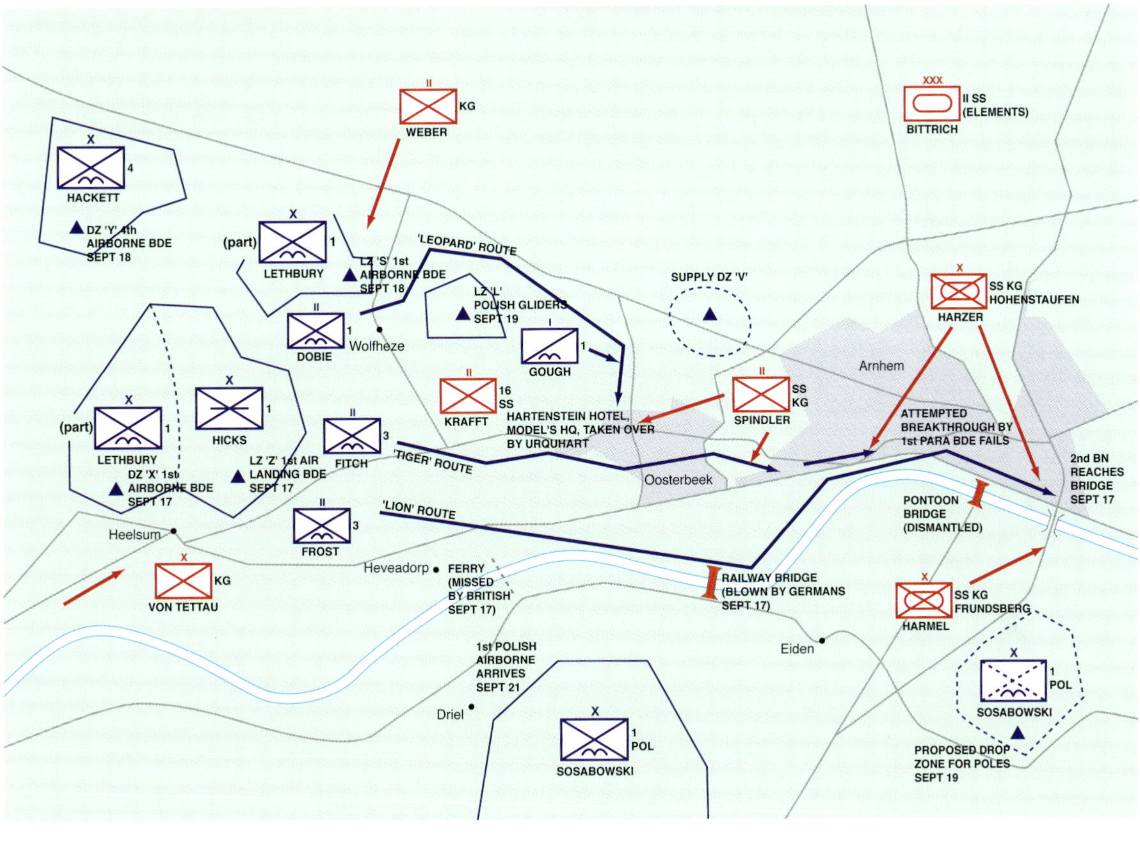

II KG WEBER

XXX II SS (ELEMENTS) BITTRICH

X 4 HACKETT

▲ DZ 'Y', 4th AIRBORNE BDE SEPT 18

X 1 (part) LETHBURY

'LEOPARD' ROUTE

▲ LZ 'S' 1st AIRBORNE BDE SEPT 18

LZ 'L' POLISH GLIDERS SEPT 19 ▲

SUPPLY DZ 'V' ▲

X SS KG HOHENSTAUFEN HARZER

II DOBIE ● Wolfheze

I 1 GOUGH

Arnhem

X 1 (part) LETHBURY

X 1 HICKS

II 3 FITCH

II 16 SS KRAFFT

HARTENSTEIN HOTEL, MODEL'S HQ, TAKEN OVER BY URQUHART

II SS KG SPINDLER

ATTEMPTED BREAKTHROUGH BY 1st PARA BDE FAILS

2nd BN REACHES BRIDGE SEPT 17

▲ DZ 'X' 1st AIRBORNE BDE SEPT 17

▲ LZ 'Z' 1st AIR LANDING BDE SEPT 17

'TIGER' ROUTE

Oosterbeek

'LION' ROUTE

II 3 FROST

PONTOON BRIDGE (DISMANTLED)

Heelsum ×

Heveadorp ● FERRY (MISSED BY BRITISH) SEPT 17

RAILWAY BRIDGE (BLOWN BY GERMANS SEPT 17)

X SS KG FRUNDSBERG HARMEL

X KG VON TETTAU

Eiden ●

1st POLISH AIRBORNE ARRIVES SEPT 21

X 1 POL SOSABOWSKI

▲

Driel ●

X 1 POL SOSABOWSKI

PROPOSED DROP ZONE FOR POLES SEPT 19

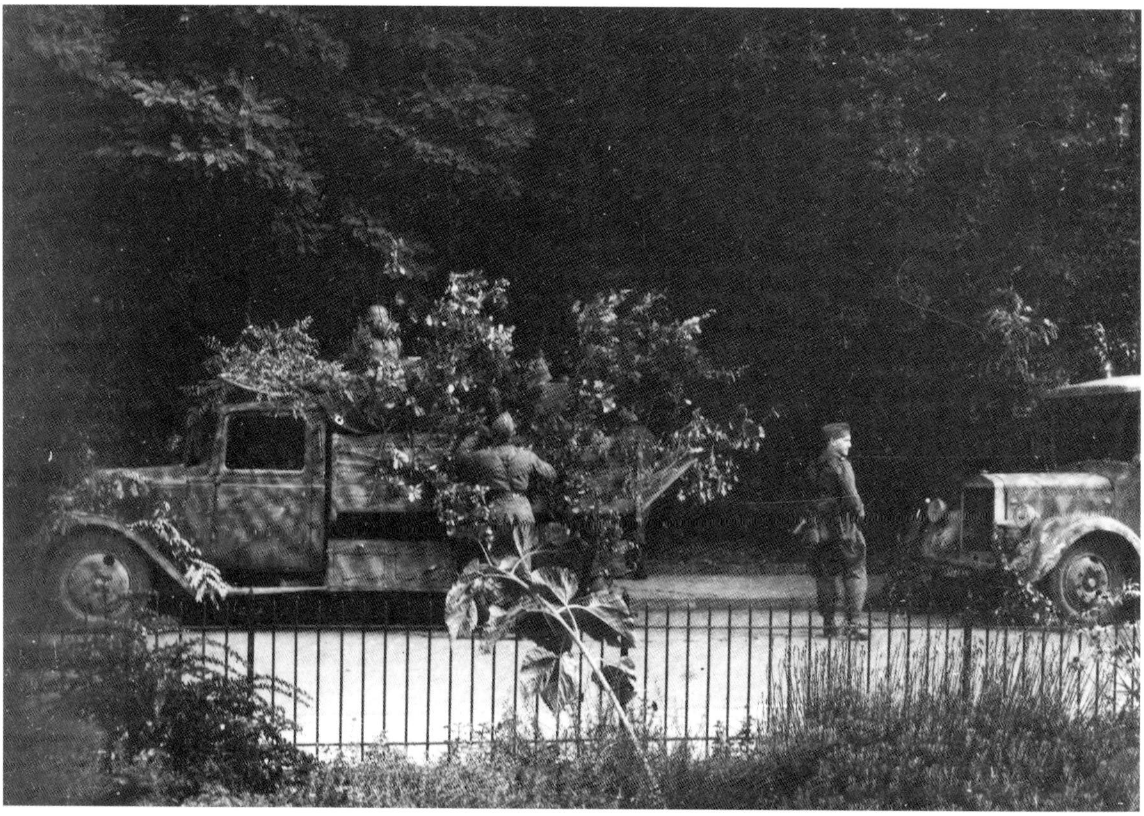

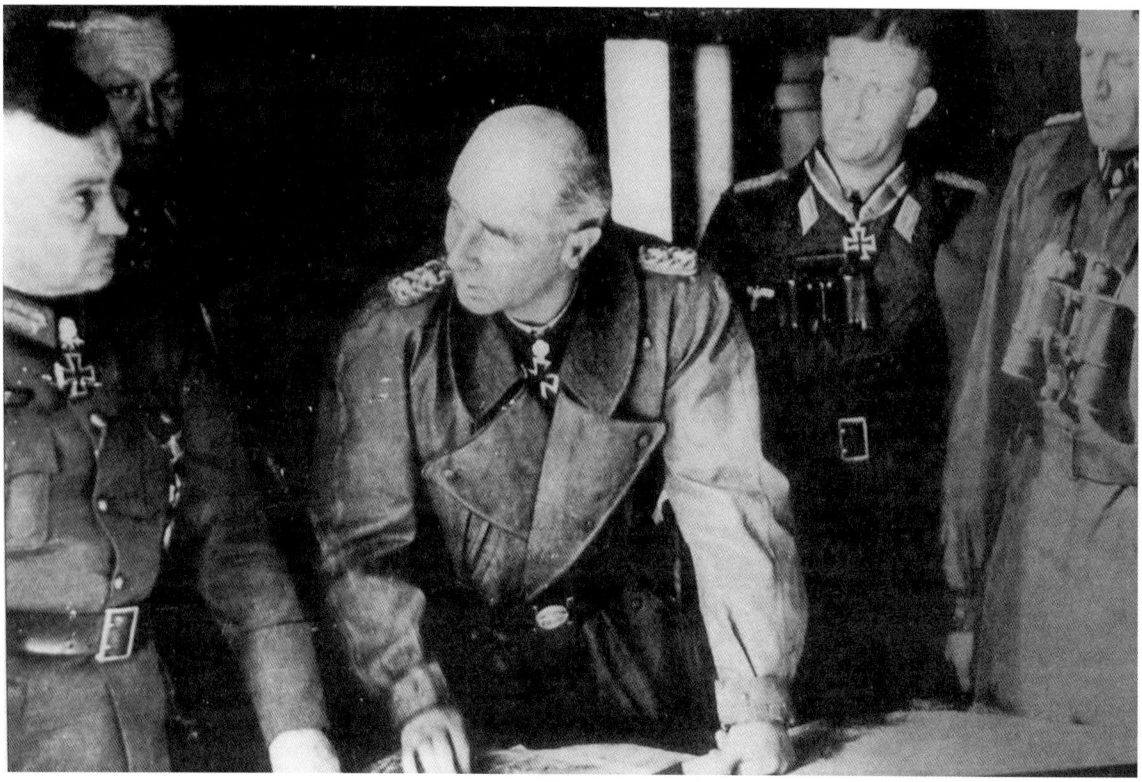

Opposite, Above, and Left:
SS PzGr Training and Replacement Bn 16, commanded by SS-Sturmbannführer Sepp Krafft, was well-placed to hold up the Paras, but it was quickly flanked and forced to move back to join KG Spindler. (**Left:** SS-Obersturmbannführer Ludwig Spindler.) This unit set up a strong blocking position on the night of September 17–18 and played a significant role in holding up the British forces.

Opposite, Below:
The architects of victory. This well-known photo appeared in *Signal* and shows a September 28 conference discussing the events of Operation Market Garden. L–R: Generalfeldmarschall Walter Model (Army Group B), Generaloberst Kurt Student (First Fallschirmjäger Army), SS-Obergruppenführer Wilhelm Bittrich (II. SS-Pz Corps, Major Hans-Peter Knaust (commander of KG Knaust, he took Arnhem bridge from the British and was awarded the Knight's Cross for his actions, receiving the award at this conference), and General der Waffen-SS Heinz Harmel (10th SS-Pz Division Frundsberg).

Above:
Looking from the dropzones to the bridge (**A**). Note the N225 Utrechtseweg (**B**) that leads into the heart of the city and railroad line (**C**) along which C Coy, 3rd Para reached the bridge.

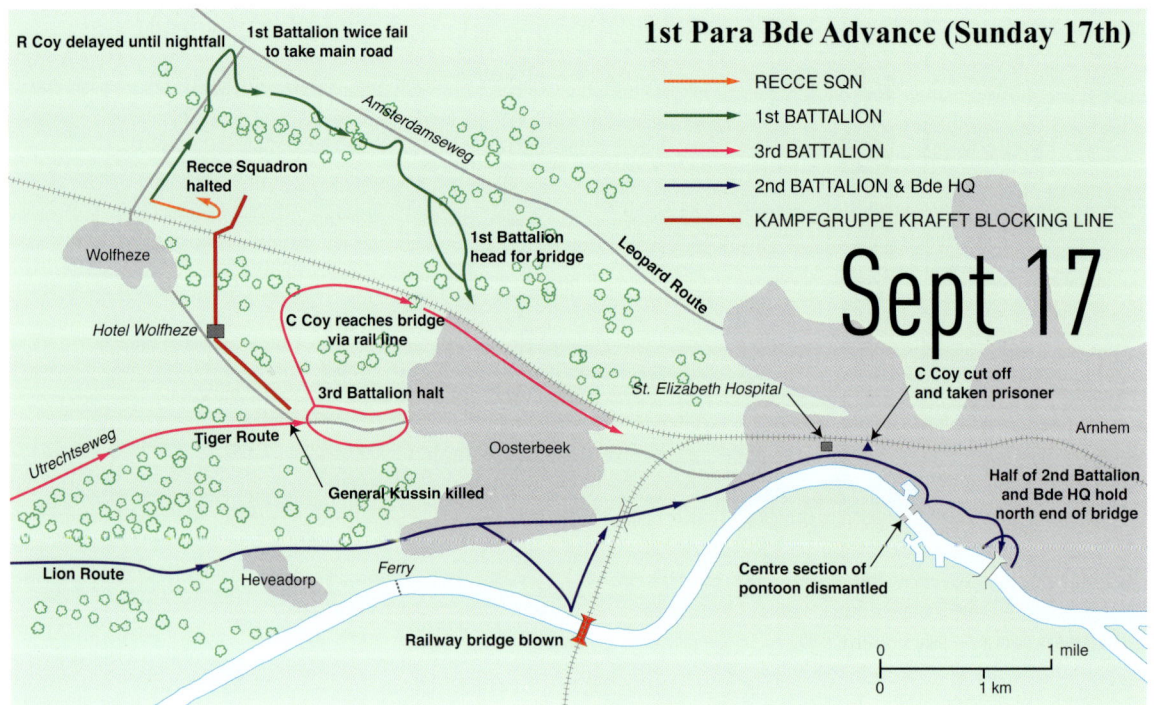

1st Para Bde Advance (Sunday 17th)

Legend:
- RECCE SQN
- 1st BATTALION
- 3rd BATTALION
- 2nd BATTALION & Bde HQ
- KAMPFGRUPPE KRAFFT BLOCKING LINE

Sept 17

Map labels:
- R Coy delayed until nightfall
- 1st Battalion twice fail to take main road
- Amsterdamseweg
- Recce Squadron halted
- Wolfheze
- Hotel Wolfheze
- 1st Battalion head for bridge
- Leopard Route
- C Coy reaches bridge via rail line
- 3rd Battalion halt
- Tiger Route
- St. Elizabeth Hospital
- C Coy cut off and taken prisoner
- Arnhem
- Utrechtseweg
- General Kussin killed
- Oosterbeek
- Half of 2nd Battalion and Bde HQ hold north end of bridge
- Lion Route
- Heveadorp
- Ferry
- Centre section of pontoon dismantled
- Railway bridge blown
- 0 ___ 1 mile
- 0 ___ 1 km

Above:
The advance of 1st Para Bn. Note the blocking line of KG Krafft which, although outflanked, did enough to slow down the British, enabling KG Spindler and 9th SS-Pz Division to put together a stronger defensive position.

Below:
The first German prisoners are marched past Wolfheze Asylum. There were some 200 kept in the tennis court area at Hartenstein on September 25–26 when the division evacuated.

At every stage in the planning of the operation it was emphasized that speed was vital. The Paras had to use surprise to achieve their objectives before the Germans could react. As we have seen, this didn't happen and nowhere is this more apparent than with 3 Para, which ended up spending the night of September 17/18 at the Hartenstein when it should have been pushing on. One of the reasons for this was that both the Divisional CO, Gen. Urquhart, and the Brigade CO,

Brig. Lathbury, were with the battalion. Urquhart had decided to chivvy on his units but became trapped with the battalion as it encountered KG Krafft. A Coy was detached to clear the woods to the north, and it was decided it was too dangerous for Urquhart to return to his HQ until A Coy had subdued the enemy. Col. Fitch (CO of 3 Para) sent C Coy to infiltrate its way to the bridge but with only B Coy left it was decided to halt on Utrechtesweg until A Coy returned.

This Page:

It's rare for generals to appear on the front line, and on D+1, September 18, Generalmajor Friedrich Kussin paid the penalty. The commander of *Feldkommandantur 642—Arnhem*—he headed out of the city toward the dropzones to assess what was happening. He met SS-Sturmbannführer Sepp Krafft and then headed back to the city down Utrechtseweg. Ambushed by B Coy, 3rd Para Bn, he and the two others in the car (his driver and batman) were killed.

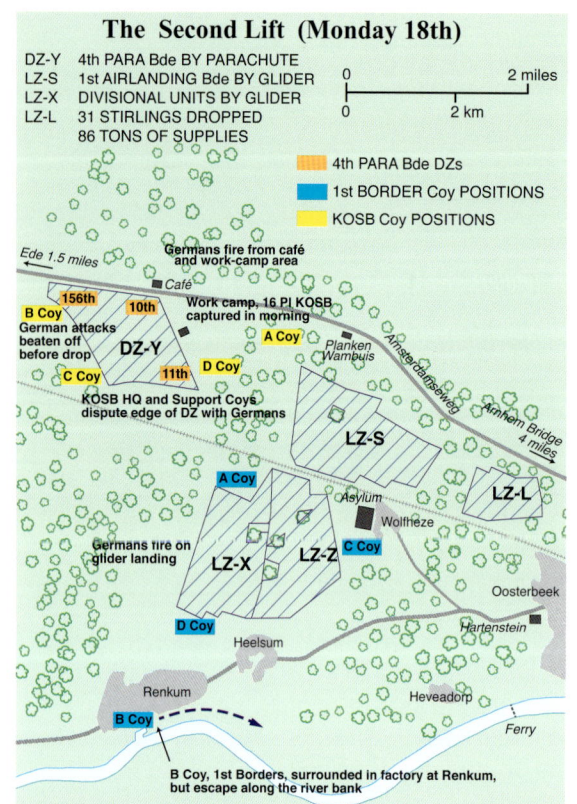

The Second Lift (Monday 18th)

DZ-Y 4th PARA Bde BY PARACHUTE
LZ-S 1st AIRLANDING Bde BY GLIDER
LZ-X DIVISIONAL UNITS BY GLIDER
LZ-L 31 STIRLINGS DROPPED
86 TONS OF SUPPLIES

0 ——————— 2 miles
0 ——————— 2 km

4th PARA Bde DZs
1st BORDER Coy POSITIONS
KOSB Coy POSITIONS

Ede 1.5 miles

Germans fire from café and work-camp area

Café

156th 10th
B Coy

Work camp, 16 PI KOSB captured in morning

German attacks beaten off before drop

DZ-Y

A Coy

Planken Wambuis

C Coy 11th

D Coy

KOSB HQ and Support Coys dispute edge of DZ with Germans

Amsterdamseweg

Arnhem Bridge 4 miles

LZ-S

A Coy

LZ-L

Germans fire on glider landing

Asylum

Wolfheze

LZ-X LZ-Z

C Coy

Oosterbeek

D Coy

Heelsum

Hartenstein

Renkum

Heveadorp

B Coy

Ferry

B Coy, 1st Borders, surrounded in factory at Renkum, but escape along the river bank

Sept 18

Above and Top right:

The Second Lift—with protection in place 4th Para Bde dropped at Ginkel Heath. There was resistance from the SS-Wachbattalion III "Nordwest," part of KG Von Tettau, who fired on the Paras as they landed (**Top right**). Total losses were surprisingly low, however: 32 men on the DZ.

Center right:

Airborne Memorial Wolfheze. "In memory of the units of the 1st British Airborne Division and of the 1st Polish Independent Parachute Brigade Group which landed in this vicinity on 17, 18 and 19 September 1944. From here they advanced in the direction of Arnhem to seize the road-bridge as part of Operation Market Garden. This battle of Arnhem lasted from 17 to 26 September 1944."

With Urquhart and Lathbury on the Utrechtesweg, "Pip" Hicks took command of the division at about 09:15 on the 18th. He immediately started to organize reinforcements for the 1st and 3rd Para Bns heading into the city—the 1st having sidestepped south from its original northerly route after coming into contact with more heavily armed German forces. Hicks immediately sent the 2nd Staffords (about 60 percent of the unit: the rest was due to land that morning) into the city. Later he ordered that the 4th Para Bde's 11th Battalion should follow on immediately after its delayed landing.

In the town, 3rd Para left its night position at 04:30 and made steady progress through Osterbeek into Arnhem, advancing to within a mile of the bridge before light opposition and sniping led to the fragmentation of the column. As the most advanced units halted to wait for those that had become detached, German armor and patrols kept them pinned down in nearby houses. The missing units did not reappear (they fell in with 1st Para). Col. Fitch decided to extract what was left of his command from the houses and try to find an alternative route to the bridge. Urquhart and Lathbury tried to return to Div HQ, but Lathbury was wounded and they were pinned down again.

continued on page 35

Left:
Prisoners from KG Krafft watched over by Glider Pilot Regiment guard.

Above right and Center right:
"Pip" Hicks, commander of 1st Airlanding Bde at his HQ in No 10 Duitsekampweg (seen today) before he took command of the division in the absence of Gen. Urquhart and Brig. Lathbury. They had both been caught up in the fighting, ending up in the Hartenstein Hotel along with 3rd Bn's B Coy who were awaiting A Coy before moving off toward the bridge.

Below right:
Men of the 1st Para Recce Sqn in defensive positions on Duitsekampweg. Note the PIAT. Capt. Robert Cain of The Royal Northumberland Fusiliers was awarded the VC for his bravery with a PIAT. The citation reads, "On September 20 a Tiger tank approached the area held by his company and Major Cain went out alone to deal with it armed with a Piat. Taking up a position he held his fire until the tank was only 20 yards away when he opened up ... Major Cain continued firing until he had ... immobilised the tank."

Opposite:
Those 2nd South Staffords who had arrived with the second lift advance to Arnhem through Oosterbeek. Battalion HQ, C, and D Companies had dropped the day before; the remainder of the battalion under Maj J.C. Cummings took off from Broadwell at around 10:30 on the 18th. The glider landings were relatively straightforward and soon the second lift was heading into Arnhem to join up with the rest of the battalion.

Above and Center right:
German prisoners are interrogated. *After the Battle* suggests the Paras are D Coy, 1st Borders.

Below right:
The battle of Arnhem caused widespread destruction in the center of the city and high numbers of civilian casualties—over 450. After the war, the survivors returned to the scene of the battle anticipating some antipathy from the population. The reality was very different: the people of Arnhem greeted them enthusiastically and have continued to do so. Regular events at the anniversary of the battle include parades by veterans, masses of reenactors, and a proliferation of memorials and museums.

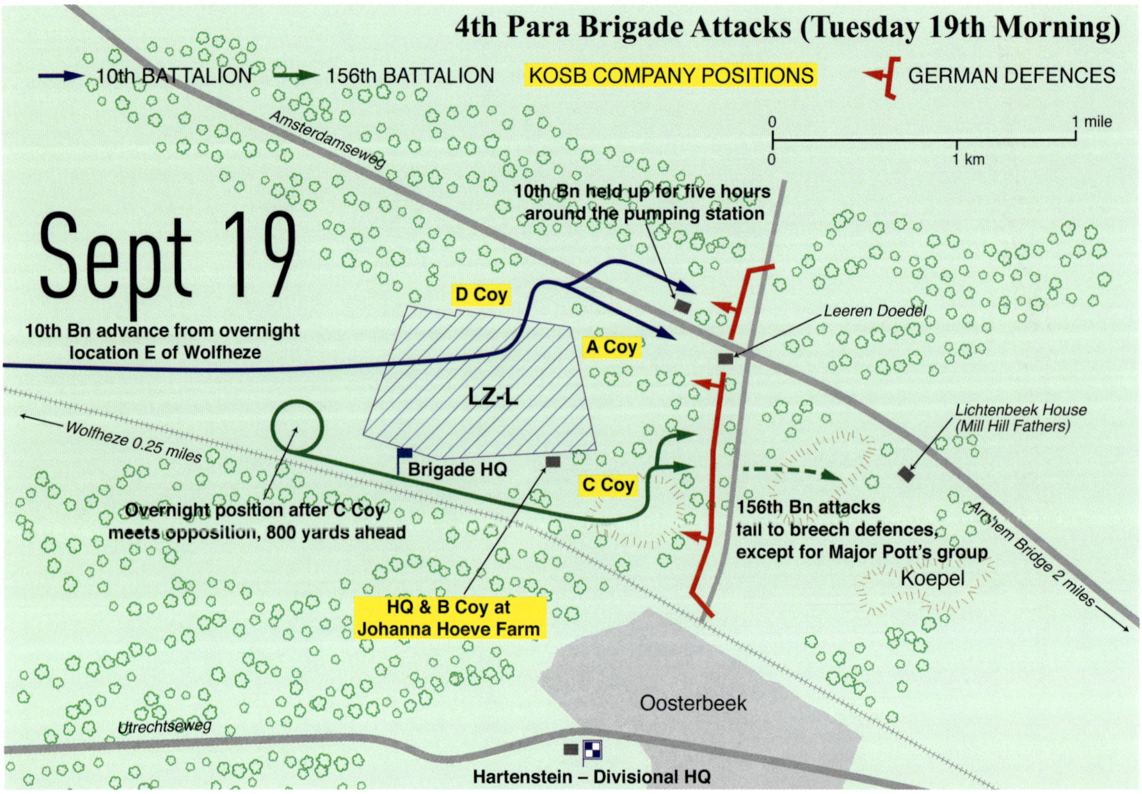

4th Para Brigade Attacks (Tuesday 19th Morning)
10th BATTALION
156th BATTALION
KOSB COMPANY POSITIONS
GERMAN DEFENCES
0
1 mile
0
1 km
Amsterdamseweg
10th Bn held up for five hours around the pumping station
Sept 19
D Coy
A Coy
Leeren Doedel
10th Bn advance from overnight location E of Wolfheze
LZ-L
Wolfheze 0.25 miles
Lichtenbeek House (Mill Hill Fathers)
Brigade HQ
C Coy
Overnight position after C Coy meets opposition, 800 yards ahead
156th Bn attacks fail to breech defences, except for Major Pott's group
Arnhem Bridge 2 miles
Koepel
HQ & B Coy at Johanna Hoeve Farm
Oosterbeek
Utrechtseweg
Hartenstein – Divisional HQ

Above:
Krafft's men had retreated along the railroad and joined KG Spindler's defensive line along Dreyenseweg. This line held the northerly attacks but had been bypassed to the south, allowing 1st and 3rd Para Bns to advance towards Frost's beleaguered force at the bridge.

GERMAN DEFENCES
3rd Bn, EARLY MORNING
1st Bn ATTACK, 04:00
3rd Bn SUPPORT 1st, 04:00
2nd SOUTH STAFFS, 04:30
1st Para Bde Final Attacks (Tuesday 19th)
0
500 yds
0
500 m
Heijenoord Diependal
Prison
St. Elizabeth Hospital
Museum
11th Bn overwhelmed while moving on Heijenoord-Diependal, 14:30
Den Brink
Utrechtseweg
C Coy, 2nd South Staffs take Den Brink, but driven off, 13:30
Onderlangs
Utrechtseweg
Pavilion
500 men withdraw to Oosterbeek in the afternoon
Harbour
River Rhine
Brickworks
Arnhem Bridge 0.5 miles
Pontoon

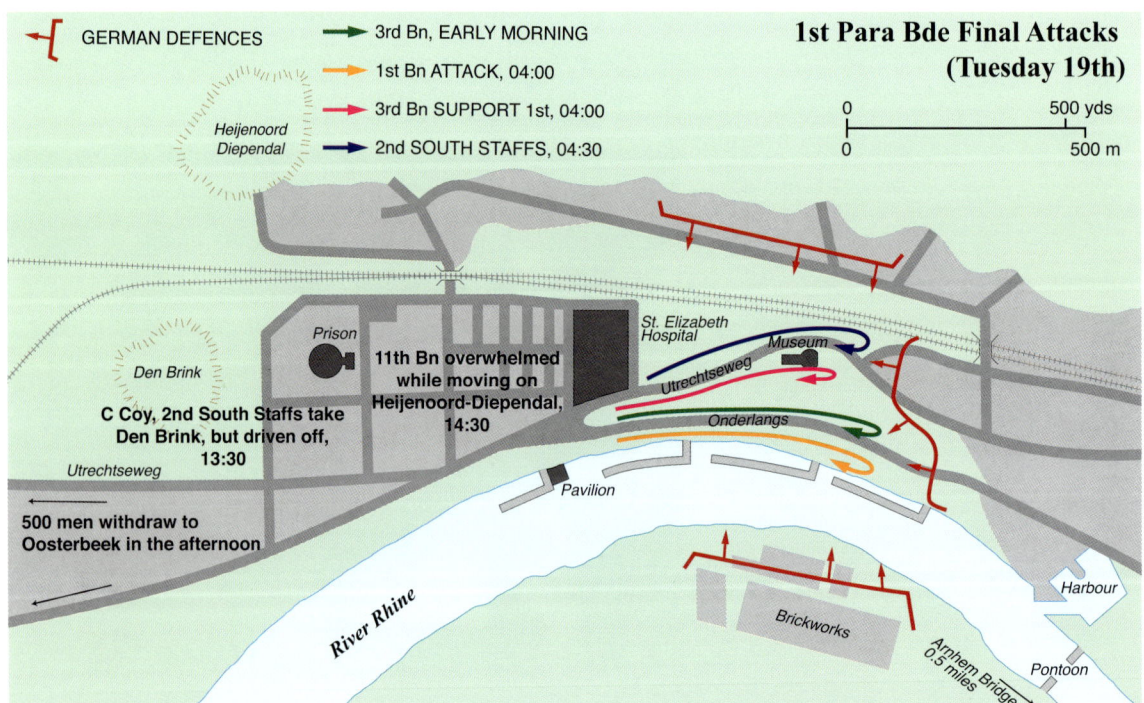

With 3rd Para Bn closest to the bridge but pinned down—along with the Divisional and Brigade COs—the advance of 1st Para became crucial. It had met up with those elements of 3rd Para that had become detached when the unit ran into opposition, and together they advanced along the same route as 3rd Para, trying to sidestep German forces and avoid any confrontation in order to get to the bridge.

Unfortunately, this tactic had to end when they reached the river and so they fought their way onward, losing men as they did: 23 in total on the Monday. In the end the defense proved too strong and they took shelter alongside 3rd Para in houses near St. Elisabeth's Hospital and awaited reinforcements. There they were joined by around 400 men of 2nd South Staffords and reinforcements from the second lift that had finally arrived: the 11th Battalion and the rest of the 2nd Staffs by about midnight.

Tuesday, September 19, was the crucial day of the battle. Lt-Col. Dobie, OC 1st Para, had planned an attack to start at 04:00 on a two-battalion front, with 1st Para and the South Staffs advancing and 11th Para following. As they moved off they ran into Lt-Col. Fitch's 3rd Para who had just tried to achieve the same objective but had been beaten back. His unit joined them to try again. It was not an auspicious start. Up ahead, units from 9th SS-Pz Div were ranged along the railroad line to the north, in a blocking line to the east, and in the brickworks on the south bank. These units had heavy weapons, armor, and plenty of ammo. The Paras didn't.

Two hours later, around 06:30, the attack was over. Faced by heavyweight opposition the Paras had cleared some of the closer German positions, but the attack had faltered. Fitch had been killed by mortar fire and the survivors had taken shelter in adjoining houses where they were cut off, winkled out, or tried to make their way back to their lines. The South Staffs fared no better, and were forced to take shelter from the StuGs that dominated the battlefield once the British had run out of PIAT bombs. They ended up in the Gemeentemuseum and most surrendered.

11th Para had been ready to support them and had taken up a good position ready for a flank attack when it received orders from Maj-Gen. Urquhart, by now back at Div HQ, not to take part but to take the high ground (the Heijenoord-Diependal feature). As they prepared to do so they were caught in the open by tanks and mortars and only around 150 men escaped.

Top and Above:
Luftwaffe photographers Jacobsen and Wenzel took the classic German views of the battle starting on the morning of the 19th. Here, a 2cm Flak gun on Jansbuitensingel.

Opposite, below:
The Paras who had bypassed KG Spindler made a final thrust towards the bridge on the afternoon of the 19th in what Kershaw likened to the Charge of the Light Brigade. They were decimated by fire from three sides, including the remnants of the SS-Aufklärungs Abt 9 across the river.

Opposite, above and below:
A wet comparison between Nieuwe Plain today and in September 1944 as the StuGs of Sturmgeschütz-Brigade 280 head towards the fighting.

Above right:
So near and yet so far ... it's only half a mile to the bridge, but it was around here on Onderlangs that the morning attack by 1st and 3rd Para Bns broke. Here, Panzergrenadiers and the detritus of the battle.

Below right and Below:
SdKfz 10/4 2cm Flak gun pointing towards the action. The comparison shot shows the changes to Arnhem in the intervening years—here caused by the Nelson Mandela Bridge.

Above:

The StuGs and accompanying infantry move past the Gemeentemuseum in which the Battalion HQ and HQ Coy of A Coy, South Staffords, had made their last stand.

Left and Below:

This plaque remembers the resistance of Maj. Dover's C Coy. Today's No 72—it was No 38 in 1944—is named Airborne House.

Above:
Surrendering men of the South Staffords. They had been obliterated in "the Hollow" by the close-range fire of the StuGs.

Left:
Today, many of the fine houses on Utrechtseweg have been replaced by blocks of flats.

Top and Above:

This photo was taken on Onderlangs, the riverside road, after the fighting had died down on the 19th. There's some debate about the identity of the helmeted man on the StuG. *After the Battle* identifies it as platoon commander Oberwachtmeister Josef Mathes, who died that day in the fighting and was awarded the Knight's Cross posthumously. Elsewhere, he is identified as the commander of 3./StuG-Brigade 280, Major Kurt Kühme. By this time the StuGs had been instrumental in beating back the Paras, whose survivors were retreating to Oosterbeek.

Above and Left:
German troops
advance towards
Oosterbook passing a
tramcar caught up in
the fighting. The photo
taken in 2017 shows
a modern house on
the site of the older
building.

Above:

Generalleutnant Hans Von Tettau's Kampfgruppe was an ad-hoc arrangement of units that attacked the landing grounds from the west. By the 19th, the force included three battalions from the SS NCO School Arnheim, Tank Company 224's six Renaults which were quickly KO'd, Regiment Knoche's three Char B1 flamethrowers, S.S.-Battalion Eberwein, and S.S.-Wach (Surveillance) Battalion 3 under Sturmbannführer Paul Anton Helle. Here troops from KG von Tettau cycle across Gingkel Heath towards the British positions.

Right:

German troops attach a 3.7cm PaK 35/36 AT gun to the back of a British jeep abandoned by 250th Airborne Lt Coy, RASC.

With the battle to reach the bridge lost, the remnants of the four battalions who had attempted the attack—the 1st, 3rd, and 11th Paras and the 2nd South Staffords—streamed back towards the comparative safety of Oosterbeek, being joined on the way by stragglers and other units. They left behind around 120 dead and 1,700 taken prisoner.

Further west, 4th Para Bde, which had landed on Monday, began a series of attacks on KG Spindler in an attempt to force its way through to the bridge along the higher ground. First 156th Para Bn attacked towards the Koepel high ground. KG Spindler was well-equipped with armored cars and StuGs and was well-sited on high ground. The attack failed and the German line held but for an incursion by Maj. John Pott, who broke through but in insufficient numbers to achieve a foothold.

North of the 156th Para attack 10th Para was enduring a five-hour battle near the pumping station. But unlike the 156th's, the 10th came away with twelve dead and no progress.

With the threat of German forces attacking the brigade from the rear, the order was given to retreat to Oosterbeek. 10th Para and the survivors of 156th Para made their way there continually harassed by the enemy. Elements of 156th Para and the Brigade HQ were pinned down in "the Hollow" before charging out with fixed bayonets. They reached the perimeter on Wednesday afternoon, raising the number of 4th Para Bde "survivors" to around 500.

Below:
Stirlings drop supplies on the 19th. In spite of the bravery of the aircrews, most of the supplies dropped by the resupply missions did not reach the Paras—under 10 percent of the 106 tons dropped. One of the problems with this was that few of the 110 radio batteries that were needed every day to cover damage and attrition reached the Paras. In an intelligent and informative article about the issues with radio communication facing 1st Airborne, Maj. John W. Greenacre argues strongly, however, that this was not the major factor in the Paras' communications problems. Indeed, he shows that the public perception that these problems were down to "the wrong crystals" and "inadequacies of their radio equipment" are also demonstrably incorrrect. He blames "Procedural errors and mishaps combined with poor timing and in some cases just bad luck" and not the equipment.

The Bridge

Above and Opposite, Above: The area that Frost's force held September 17–21 has changed out of all recognition since 1944. See also the diagram on p46.

Around 750 men reached Arnhem Bridge, the bulk of them from Lt-Col. John Frost's 2nd Para and 1st Para Bde's HQ. They had got there by moving quickly along the southern—Lion—route. A 2nd Para detachment had gone to recce the railroad bridge and, if possible, cross it and advance to the road bridge along the southern bank. They found the bridge but it was blown as they advanced onto it. Frost's lead platoon reached the road bridge at about 20:00 on the 17th. Immediately, the Paras deployed into 27 buildings surrounding the bridge. Heavy armament included five 6-pounder AT guns, mortars, flamethrowers, and PIATs. The major deficiency was lack of RAMC surgical teams.

The first major attack the defenders had to repulse was made by Hauptsturmführer Viktor Gräbner and his SS-Pz Aufklärungs-abteilung 9, which carried in twenty-two armored cars and APCs. It took place on Monday, September 18. The attack was beaten off and its commander killed with assistance from artillery at Oosterbeek.

Tuesday and Wednesday saw continuous fighting. Attempts to blow one of the spans saw the Germans lay explosives and the defenders fight to remove the fuzes. It was during this action that Lt. Jack Grayburn was killed: he was awarded a posthumous VC.

The defenders surrendered at 05:00 on Thursday September 21. 81 of them were dead or would die of wounds.

This page:
The bridge remained intact through the battle, but was bombed by 344th Bomb Group on October 7. Temporarily replaced by Bailey bridges, in 1948 a new bridge was erected, almost identical to the original. It was named the John Frostbrug in 1977.

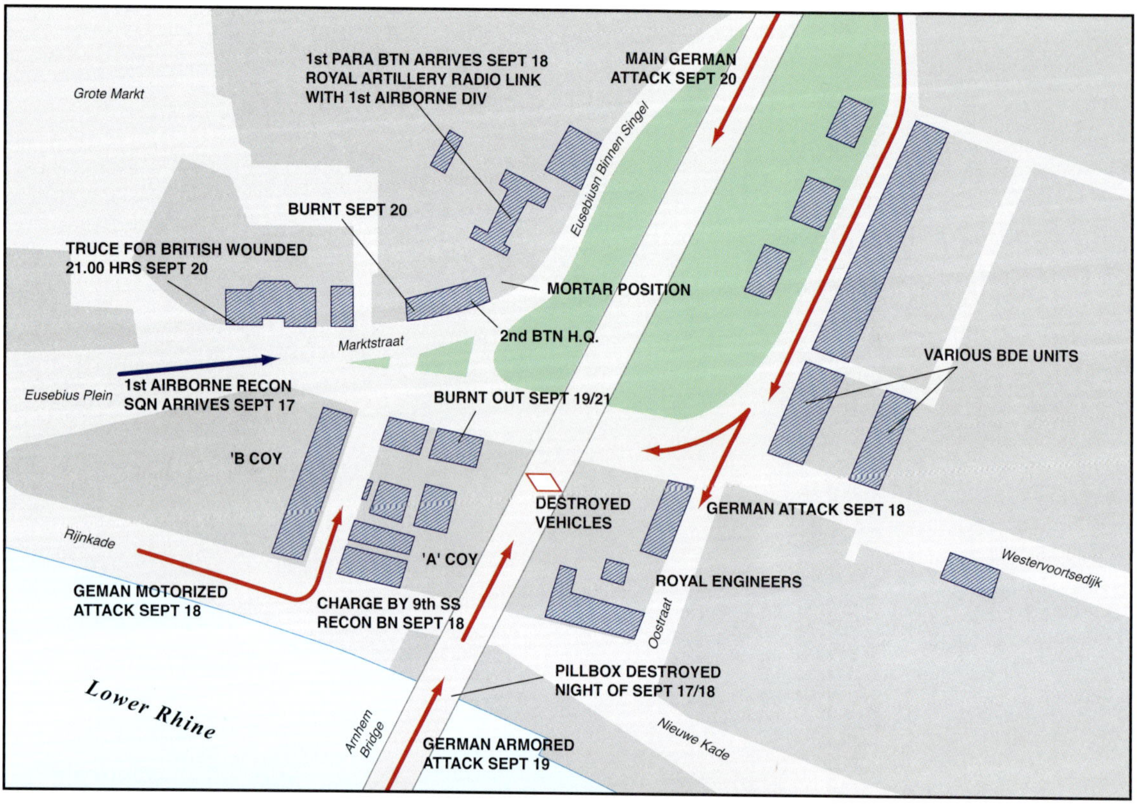

Map labels:

Grote Markt

1st PARA BTN ARRIVES SEPT 18
ROYAL ARTILLERY RADIO LINK
WITH 1st AIRBORNE DIV

MAIN GERMAN
ATTACK SEPT 20

Eusebiusn Binnen Singel

BURNT SEPT 20

TRUCE FOR BRITISH WOUNDED
21.00 HRS SEPT 20

MORTAR POSITION

2nd BTN H.Q.

Marktstraat

VARIOUS BDE UNITS

Eusebius Plein

1st AIRBORNE RECON
SQN ARRIVES SEPT 17

BURNT OUT SEPT 19/21

'B COY

DESTROYED
VEHICLES

GERMAN ATTACK SEPT 18

Rijnkade

Westervoortsedijk

'A' COY

ROYAL ENGINEERS

GEMAN MOTORIZED
ATTACK SEPT 18

CHARGE BY 9th SS
RECON BN SEPT 18

Oostraat

PILLBOX DESTROYED
NIGHT OF SEPT 17/18

Lower Rhine

Arnhem Bridge

Nieuwe Kade

GERMAN ARMORED
ATTACK SEPT 19

Above:
The locations of the force around the north end of the bridge.

Left and Above right:
On September 18, at around 09:00, the 9th SS-Pz Div Hohenstaufen's Recce Bn returned from Nijmegen and tried to cross the bridge. It was stopped with heavy losses—over 70 were killed including commanding officer Hauptsturmführer Viktor Gräbner—in a two-hour battle.

Right:
Airborne reenactors with carrier pigeons. During the battle one pigeon, William of Orange, flew 260 miles back to Knutsford, Cheshire, in just 4hr 25min with the message that they were in dire need of support.

Above:

The defenders around the bridge had held for three days and nine hours before they surrendered at around 05:00 on the 21st. Of the 750 men, 81 were dead and many were wounded. These are Sappers Grier and Robb pictured by a Luftwaffe photographer.

Above left and Left:

These PzKpfw IVs of a training company from Pz-Ersatz-Regt 6, part of KG Knaust, were KO'd in the attack on the 18th.

Opposite, Above, Center left and Right:

St. Eusebius' church was almost completely destroyed in the fighting. It was rebuilt 1946–64. The Anna Chapel has a stained-glass window, whose burning city and text relate to the destruction of Arnhem. The Raads Chapel has 19 bronze parachutists descending from the ceiling—an artwork by Simona Vergani.

Below right:

A plaque remembers the site of Frost's HQ on the corner of Prinsenhof and Oranjewachtstraat.

HIER
WAS IN SEPTEMBER
1944
HET HOOFDKWARTIER
VAN LUITENANT-KOLONEL
JOHN D. FROST
HERE WAS IN
SEPTEMBER
1944
THE HEAD-
QUARTERS
OF
LIEUTENANT-
COLONEL
JOHN D.
FROST

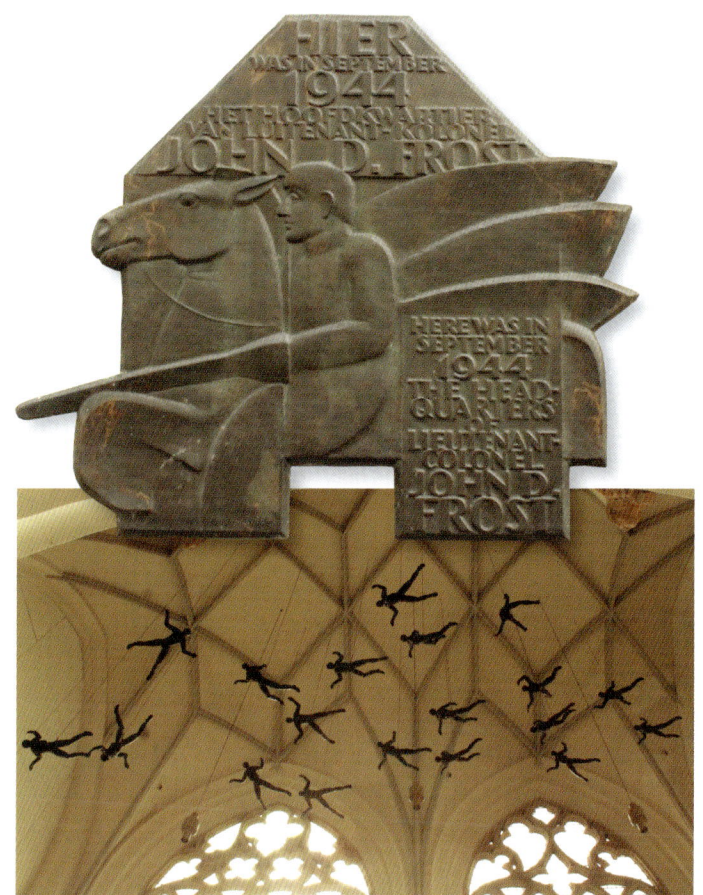

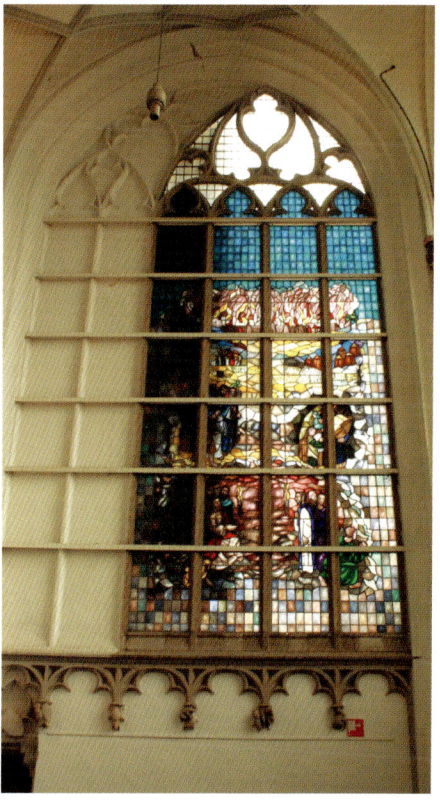

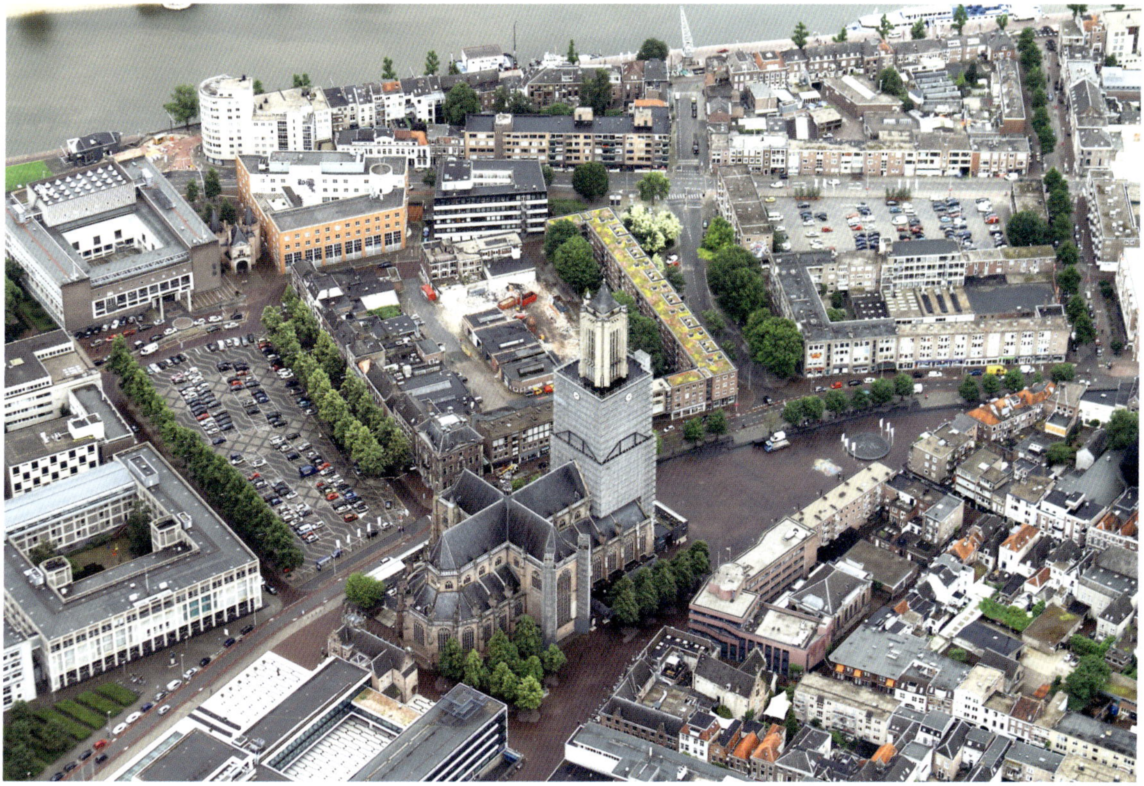

Oosterbeek

Below and Below right:
An abandoned vehicle outside Oosterbeek—Lonsdale—Church on Benendendorpsweg. The chuch spire, severely weakened by the shell fire and mortar hits, finally collapsed on the last day of battle. It was here that Maj. Richard Lonsdale made his famous speech, "...We must fight for our lives and stick together. We've fought the Germans before—in North Africa, Sicily, Italy. They weren't good enough for us then, and they're bloody well not good enough for us now. ..."

THE REVERSALS OF TUESDAY saw the end of 1st Airborne's offensive capabilities. Now it was a matter of survival—and as the survivors of the fighting moved back towards Oosterbeek they formed into a perimeter which was held until the night of the 25/26th.

As with almost every detail of the operation, there is debate about the positioning of this perimeter—that it was centered on the Hartenstein Hotel rather than the commanding Westerbouwing heights where the ferry had plied its trade.

Its positioning was mainly by chance. As men appeared they were fitted into a defensive structure that was about three miles long. 1st Bn, the Border Regt defended most of the western perimeter, including the Westerbouwing Heights; the 21st Ind Para Bn and 7 KOSB the north—although this area was given up as Urquhart shortened his lines on the 21st; the northeastern edge was covered by the remnants of 4th Para Bde (10th and 156th Battalion) and 1 Recce Sqn; and finally, the east edge was covered by the 2nd Staffords and Thompson Force—renamed Lonsdale

Force after Lt-Col. Thomas's injury on the 21st. It was in advance of this area, while covering the retreat, that Jack Baskeyville (see page 6) and Robert Cain (see page 31) won their VCs.

There were around 3,600 men inside the perimeter and 2,500 civilians. For five days they endured relentless shelling, mortaring, sniping, and regular attacks using armor and flamethrowers.

On the plus side, on Thursday, September 21, contact was made with artillery units in Nijmegen and from then the beleaguered defenders could call upon artillery support.

There were many attempts to resupply the defenders by air, but the cost was great. On the 20th 17 aircraft didn't return home; on the 21st 28 aircraft were shot down; on the 23rd 11—and very few of the supplies on these or other days reached their intended recipients.

It was only a matter of time before the perimeter became untenable. It was essential either to reinforce the defenders with men, munitions, and heavy weapons or to evacuate them.

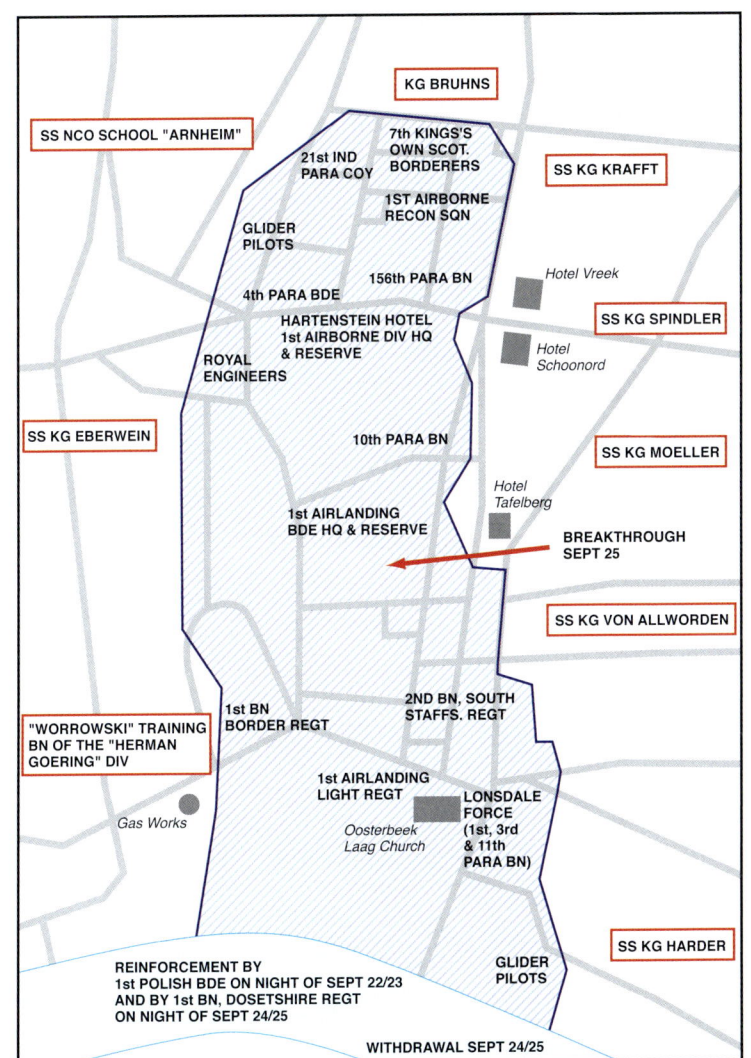

Map labels:

SS NCO SCHOOL "ARNHEIM"

KG BRUHNS

7th KINGS'S OWN SCOT. BORDERERS

21st IND PARA COY

1ST AIRBORNE RECON SQN

SS KG KRAFFT

GLIDER PILOTS

156th PARA BN

Hotel Vreek

4th PARA BDE

HARTENSTEIN HOTEL 1st AIRBORNE DIV HQ & RESERVE

SS KG SPINDLER

Hotel Schoonord

ROYAL ENGINEERS

SS KG EBERWEIN

10th PARA BN

SS KG MOELLER

Hotel Tafelberg

1st AIRLANDING BDE HQ & RESERVE

BREAKTHROUGH SEPT 25

SS KG VON ALLWORDEN

"WORROWSKI" TRAINING BN OF THE "HERMAN GOERING" DIV

1st BN BORDER REGT

2ND BN, SOUTH STAFFS. REGT

Gas Works

1st AIRLANDING LIGHT REGT

Oosterbeek Laag Church

LONSDALE FORCE (1st, 3rd & 11th PARA BN)

SS KG HARDER

GLIDER PILOTS

REINFORCEMENT BY 1st POLISH BDE ON NIGHT OF SEPT 22/23 AND BY 1st BN, DOSETSHIRE REGT ON NIGHT OF SEPT 24/25

WITHDRAWAL SEPT 24/25

Left:

The Oosterbeek Perimeter. By the 20th, the Paras were constricted in a defensive perimeter, around which the battle raged for six days against a number of German Kampfgruppen, reinforced from the 24th by a company of 14 Tiger IIs of sPzAbt 506, one of which was knocked out on the 25th by a combination of artillery fired from Nijmegen and anti-tank fire from the perimeter.

Below left:

PzKpfw B2 (f) Flammenwerfer (flamethrower tank) of Panzer-Kompanie 224 KO'd by *Gallipoli II* of No. 26 AT Platoon, 1st Border Regt on September 20.

Below:

Gallipoli II engages the Char B1(f).

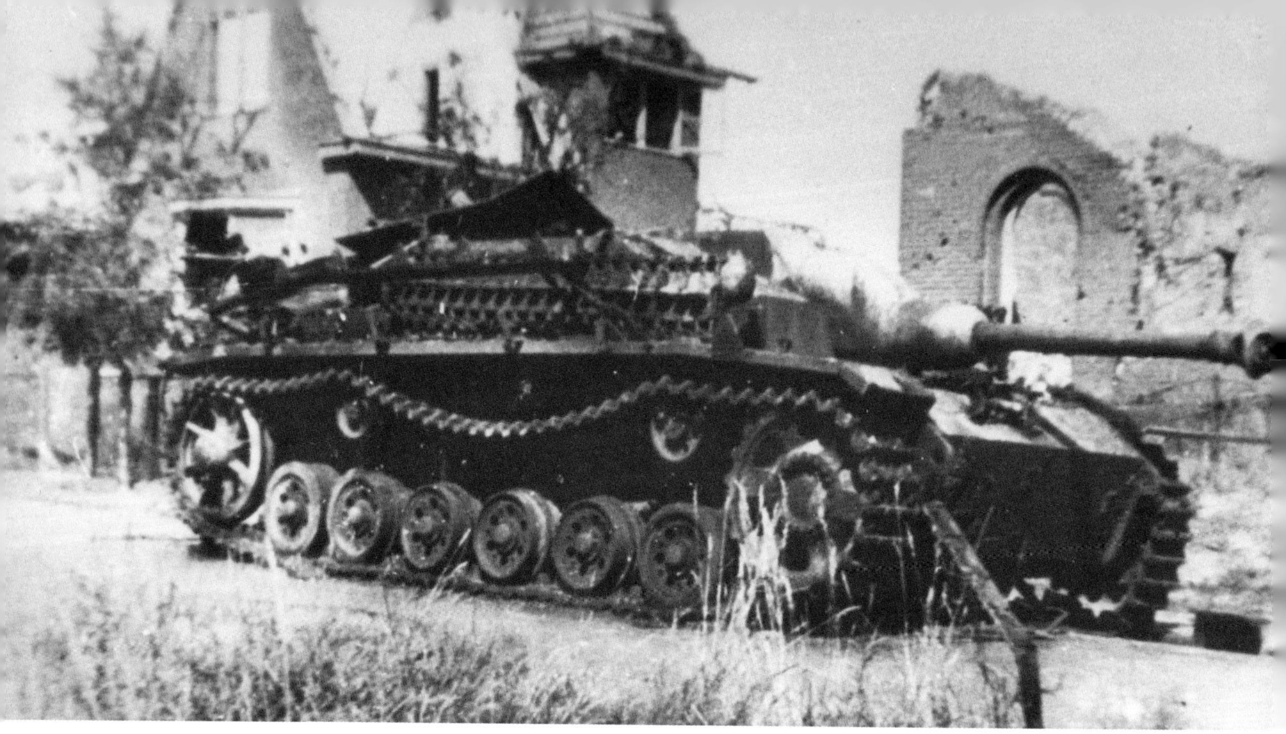

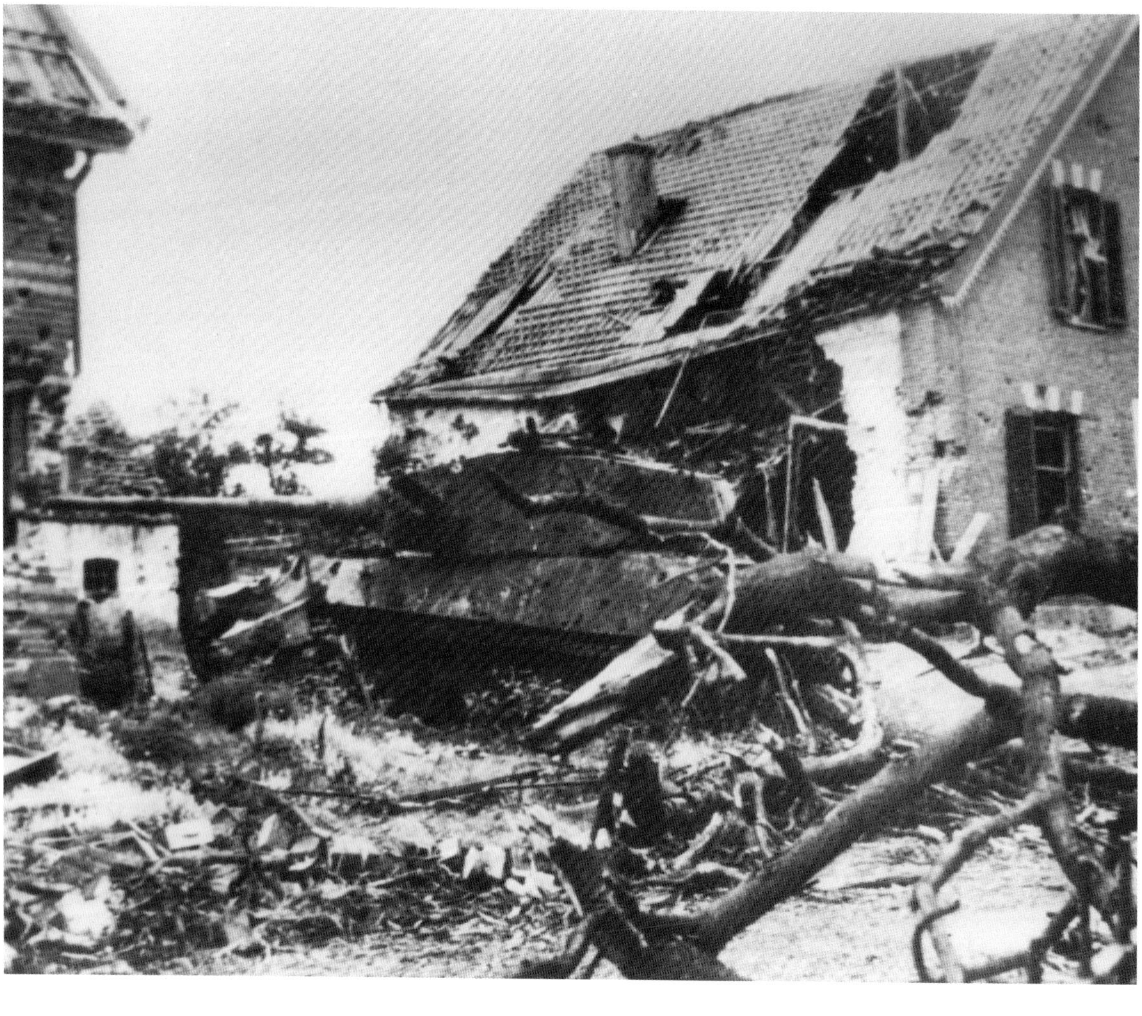

Opposite, Above:
StuG III KO'd by Maj. Robert Cain using a PIAT on Setember 21. Cain, of the Royal Northumberland Fusiliers, attached to the South Staffs, was awarded the VC for his courage and leadership during the battle.

Opposite, Below:
Tiger II of 2./sPzAbt 506 KO'd on October 25. Claimed by many people, Karel Magry suggests that it was Lt. Adrian Donaldson and Lance Bombardier Joe Dickson of C Troop, the Light Regt., who got it.

Left and Below:
2./StuG-Bde 280 had three StuH 42Gs armed with 105mm howitzers. These are seen on Weverstraat (left) and Van Eeghweg.

Above:

Aerial view of the Hartenstein showing (**1**) one of two 17pdr ATk guns preserved. (**2**) Arnhem Aircrew Memorial to the memory of the RAF, Commonwealth, and USAAF aircrew who died during the operation. It was dedicated in 2006. (**3**) Memorial "To the people of Gelderland: 50 years ago British & Polish Airborne soldiers fought here against overwhelming odds to open the way into Germany and bring the war to an early end. Instead we brought death and destruction for which you have never blamed us. This stone marks our admiration for your great courage, remembering especially the women who tended our wounded. In the long winter that followed your families risked death by hiding Allied soldiers and airmen, while members of the Resistance helped many to safety. You took us then into your homes as fugitives and friends, we took you forever into our hearts. This strong bond will continue long after we are all gone." (**4**) Sherman V *Argyle* of 4th Tp, A Sqn, 2nd Tank Regt, Canadian 5th Armd Div. This saw action in April 1945 when it assisted in clearing German troops to the west of Arnhem.

Below Left and Right:

Maj-Gen. Roy Urquhart outside the Hartenstein.

1 Arnhem Aircrew Memorial (**2** on aerial photo).

2 Firing an American M1 carbine from the front balcony of the Hartenstein. The weapon may have been acquired from one of the 12 American soldiers present at Divisional HQ during the battle.

3 Resupply drop on September 21.

4 Memorial to the people of Gelderland (**3** on aerial photo).

1 Probably taken on the 21st, this mortar team is from 23 Mortar Pl, Support Coy, 1st Border Regt.

2 Another photo of members of No. 26 AT Platoon, 1st Border Regt on September 20.

3 Pvt Joe Cunningham, one of the members of the *Gallipoli II* gun crew.

4 Walking wounded leaving the Tafelberg for St. Elisabeth Hospital under a flag of truce, September 24. This German photograph shows the wounded on the Utrechtseweg between Oosterbeek and Arnhem just before the railroad viaduct at Mariendaal.

5 The Air Despatch Memorial in Oosterbeek commemorates the 79 members of the Air Despatch Squadron who died during the battle of Arnhem.

6 The Airborne Memorial in front of the Hartenstein. Each of the sides of the column has a relief: That on the east side commemorates the landings; the south, the women of Oosterbeek and their help for the wounded; the west, the resistance and its aid to the Airborne; the north, 1st Airborne's last stand.

7–9 The Westerbouwing restaurant has a number of memorials and a superb view of the area. This (**7**) commemorates the soldiers of the Dorset Regiment; (**8**) the 1st Border Regiment; (**9**) the 4th Bn, Dorset Regiment.
7—Pimvantend/WikiCommons (CC BY-SA 3.0)

The End

Below:

The advance from Nijmegen and the last attempts to rescue the men trapped in the Oosterbeek Perimeter.

THE LAST ACT OF Operation Market Garden was as muddled as much of the rest of it. The Polish Independent Parachute Brigade was finally dropped on Thursday, September 21—two days later than planned. And the drop was not complete. Bad weather led to a general recall of the 314th and 315th Troop Carrier Groups, but problems with codes meant 73 aircraft continued while 41 turned back. Instead of 1,568 men, 1,003 arrived at Driel.

Once on the ground, things didn't get any better. There were no boats and Sosabowski was alarmed to find that the Westerbouwing Heights were no longer in British hands. There were a number of abortive attempts to get men over to Oosterbeek. (In the end around 150 made it over on Saturday, 23rd.)

At 01:00 on the 25th, 315 men of the 4th Dorsets attempted to cross the Rhine and reinforce the defenders. As Gen. Sosabowski said when told of the plan (one of his battalions was to follow the Dorsets), the crossing was in the wrong place and would be a disaster. He was correct on both counts. Thirteen died and some 200 were taken prisoner.

There was little left but to evacuate the bridgehead. On the night of September 25/26—a dreadful night, with the rain bucketing down—the evacuation was handled so successfully that the German forces only realized what was going on after Operation Berlin had evacuated around 2,400 men. A further 140 would cross the river later aided by the Dutch. 400 wounded were left behind and moved to an "Airborne Hospital" in Apeldoorn where they were tended by the British medics who had stayed with them.

The exhausted survivors were fed and watered and taken to Nijmegen where the Pagoda (a missionary college) and Training College for Catholic Girls had been prepared for their arrival. Over half the nearly 12,000 men involved in the operation from 1st Airborne Division and the Polish Brigade—around 6,500 soldiers—entered captivity.

Efforts to rescue British 1st AB Division

Right:

The Polish 1st Independent Parachute Bde became the scapegoats for the failure at Arnhem. Here (**2**) memorial to the brigade erected in memory of the Polish soldiers who were killed. (**3**) their commander, Maj-Gen Stanisław Sosabowski, is seen with "Boy" Browning. His memorial (**1**) was erected by British veterans who fought with the Poles.

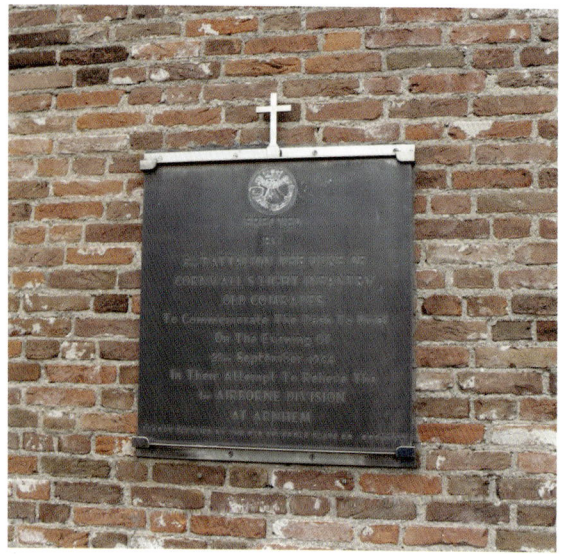

Above:

Plaque erected on the church at Driel: "Erected by 5th Battalion the Duke of Cornwall's Light Infantry ...To Commemorate The Dash To Driel On The Evening Of 22nd September 1944."

Above:

Monument to the 7th Royal Hampshire Regiment at Driel.

Below:

After crossing back over the Rhine, the survivors stacked their weapons, had a meal, and then slept! This group sorting through weapons is at the Pagoda Missionary College in Driel.

Above:
Operation Berlin—the evacuation of around 2,400 troops who had defended the Oosterbeek Perimeter—left the north bank from here on the night of September 25/26.

Right:
The memorial for the Royal Engineers (RE) and Royal Canadian Engineers (RCE) at Driel commemorates the evacuation. The boats were manned by men of the 553rd Field Company, RE and the 20th Field Company, RCE (western route) and 260th Field Company, RE with the 23rd Field Company, RCE (east).

Below:
Survivors outside the Pagoda.

17
SEPTEMBER
1944

Above:
Airborne reenactor with Bren gun.

Left:
Airborne memorial in the Berenkuil (Bearpit) roundabout above the John Frost Bridge—on an old pillar rescued from the Justice Palace.

Opposite, Above:
The German cemetery in the Netherlands is at Ysselsteyn. Martin Middlebrook quotes a signal "presumably" from II SS-Pz-Korps indicating 3,300 casualties including 1,300 dead.

Opposite, Below:
The CWGC cemetery at Oosterbeek has 1,759 dead—British and Poles—many of whom fell during the week's fighting at Arnhem.

Cemeteries

Bibliography

Traces of War (http://en.tracesofwar.com) is a fount of knowledge about memorials, fortifications, cemeteries, points of interest, awards: definitely worth checking out.

The Pegasus Archive (http://www.pegasusarchive.org) is a mine of useful information including many of the unit war diaries.

Buckley, John & Preston-Hough, Peter: *Wolverhampton Military Studies 20 Operation Market Garden*; Helion, 2016.

Frost, John: *Nearly There*; Leo Cooper, 1991.

Greenacre, Maj. John W.: "Assessing the Reasons for Failure: 1st British Airborne Division Signal Communications during Operation 'Market Garden',"; *Defence Studies*, 4:3, 283-308, DOI: 10.1080/1470243042000344777, 2004.

Holt, Maj. & Mrs.: *Operation Market-Garden*; Pen & Sword, 2012.

Hook, Patrick: *Spearhead 20 Hohenstaufen 9th SS Panzer Division*; Ian Allan Ltd, 2005.

Kershaw, Robert J.: *'It Never Snows in September'*; Ian Allan Ltd, 1994.

Kershaw, Robert J.: *A Street in Arnhem*; Ian Allan Ltd, 2014.

Margry, Karel [ed.]: *Operation Market Garden Then and Now*; After the Battle, 2002.

Middlebrook, Martin: *Arnhem 1944*; Penguin Books, 1995.

Steer, Frank: *Battleground Europe Arnhem: The Landing Grounds and Oosterbeek*; Pen & Sword, 2002.

Key to Map Symbols

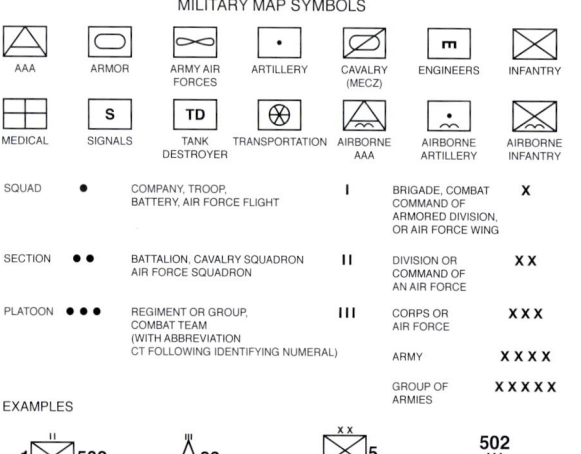

Remember September 1944 memorial on Nassaustraat, Arnhem.